DANISH SHIP DESIGN 1936-1991

The work of Kay Fisker and Kay Kørbing

BRUCE PETER

©Ferry Publications Ltd/Bruce Peter 2004
PO Box 33, Ramsey,
Isle of Man, British Isles
IM99 4LP
Tel: +33(0) 1624 898445
Fax +33(0) 1624 898449

The River Tyne as it appeared in the latter 1970s and early 1980s; the *England* manoeuvres past decaying wharfs close to the DFDS berth at Tyne Commission Quay in 1981. The ship was withdrawn at the end of the 1982 season and the riverside has since been redeveloped with yacht harbours and new housing. (Ambrose Greenway)

FOREWORD

During its 137 years of existence, DFDS has developed a reputation not only for the safety and reliability of its ships but also for offering a distinctive onboard ambience. Much credit for this reputation goes to Kay Fisker and Kay Kørbing – two highly regarded Danish architects who revolutionised passenger ship design from the 1930s onwards. The ships they designed for DFDS introduced a whole range of innovations to overnight routes – open plan public rooms, luxury cabins, cocktail bars, writing rooms, shops and even children's play areas. The short sea voyage was transformed and passengers began to sail for pleasure, rather than simply to get from A to B.

Kay Kørbing, in particular, was at the height of his powers during the 1960s and 70s, when such favourite ships as *England*, *Winston Churchill*, *Dana Regina* and *Tor Britannia* were introduced on North Sea routes. Since then, however, a great deal of change has occurred in the passenger shipping industry and DFDS has changed too. In the 1960s, the company's passenger route network had a distinctly Danish focus, converging on the ports of Esbjerg and Copenhagen – including a number of domestic routes from Aalborg and Århus.

Now, three decades later, the DFDS passenger ship division has a much wider North European focus, linking Britain with Denmark, the Netherlands and Germany and Denmark with Poland. Even so, the issue of design and image remains most important – although the design values themselves have altered over the years.

Whilst the ships designed by Kay Fisker and Kay Kørbing were essentially miniature ocean liners, primarily offering comfortable point-to-point transportation, the largest passenger ships in the DFDS fleet today are as much cruise ships as means of transportation and the philosophy behind cruising is quite distinct. Essentially, a liner or ferry takes one from A to B, often in great comfort, whereas cruising is largely about fantasy, escapism, indulgence and nostalgia.

Today, the DFDS flagship, MS *Pearl of Scandinavia*, offers passengers on the Copenhagen-Oslo route a luxurious cruise experience with an Oriental accent. Designed and built to combine the roles of cruise ship and ferry, the vessel sailed initially in the Baltic, then moved to the Far East. By the time DFDS acquired the ship, her interiors had been refitted in a South East Asian manner, which DFDS considered to be most effective and a unique attraction for a passenger ship sailing in Scandinavia. In this way, the interiors of modern short-sea passenger ships have changed from Danish modernism to embrace the aesthetics of the global leisure industry. Nevertheless, for the designers of today's passenger ships, the pioneering work of Kay Fisker and Kay Kørbing remains a benchmark for quality and comfort at sea.

DFDS would like to extend our sincere gratitude to Bruce Peter for being so enthusiastic about Danish passenger ship design and taking the effort to write this interesting book about Danish passenger ships. Also, our thanks go to Miles Cowsill and John Hendy of Ferry Publications for publishing this book.

DFDS A/S
Copenhagen, January 2004

DANISH SHIP DESIGN 1936-1991

PREFACE

My first encounter with modern Danish architecture was in the unlikely setting of North Shields on the River Tyne in England. There, tied up at the quay, amid the post-industrial dereliction of Tyneside, lay the sleek, white Danish passenger liner *England*, her streamlined black and red funnel and raked masts protruding above the terminal building. I was just a boy then and about to go on holiday to Denmark with my parents. Sailing to Denmark to meet my Danish family was an adventure and the *England* certainly made a deep impression.

Many years and many North Sea crossings later, I finally met the architect of the *England* and so many other fine Danish passenger liners and ferries – Kay Kørbing – at his beautiful farmhouse near Grenå. It was a fascinating and enjoyable day. Obviously Kay was delighted to meet someone who had appreciated his design work and he spoke with youthful

The *Dana Regina* is seen shortly after her introduction on the Esbjerg-Harwich route in 1974. (DFDS)

enthusiasm, which belied his years. Even in retirement, his passion for architecture was undiminished. He then began to recall his career – from the early-1930s when he laboured as a bricklayer on Copenhagen building sites, to his years in the hothouse atmosphere of the Royal Academy's Architecture School at Charlottenborg, surrounded by such great talents as Kay Fisker, Vilhelm Lauritzen and Jørn Utzon.[1] As Professor of Architecture there, Fisker was Kørbing's mentor and the two shared a similar attitude to design and to the value of good detailing. Then we started to talk about the ships – those beautiful creations that intrigue and enchant us all.

Between them, Kay Fisker and Kay Kørbing's ship designs span the most innovative era in twentieth century Danish architecture – from the mid-1930s until the 1990s – when there was constant design evolution. Study of this relentless and rigorous development process not only reveals a great deal about changing attitudes to modern design in Denmark and beyond, but also demonstrates an evolving relationship between the modern movement in architecture and industrial design on the one hand and developments in naval architecture on the other.

Modernism demanded a more integrated approach to naval architecture; passenger ship interiors in particular were no longer to be considered as an afterthought in the design process. Rather, Fisker and Kørbing determined that they were the starting point, and so their contributions to the design of the ships described in this book usually went well beyond mere interior outfitting and affected the structural organisation and external styling as well. Nearly every aspect involved original thought and creativity – from the desire for open planning to the design of the lighting, individual pieces of furniture, door handles and the commissioning of artworks. Even fabrics for curtains and upholstery were specially made. The overall effect was of great coherence and demonstrated a persuasive clarity of intent. Yet, achieving these most visible advances required much hidden technical innovation, research and unprecedented co-operation between architects, naval architects, shipyards and shipowners.

Unlike the staid permanence of architecture on *terra firma*, ships are ephemeral and they blur the distinction between what has traditionally been thought of as architectural – enshrined and located – and the engineered – what is glimpsed in motion, passing at speed. They have inevitably finite careers and those that avoid accident or disaster are scrapped. While one can still enjoy strolling through the parkland setting of Århus University, designed by Fisker, C.F. Møller and others, or browsing in the Illums Bolighus store in Copenhagen, by Kørbing, almost all of their ships have long-since been scrapped or altered out of all recognition.

In line with the ever more self-referential nature of the global hospitality and leisure industry, passenger shipping has become increasingly divorced from the mainstream of architectural discourse. The retro 'kitsch' of a cruise liner interior in the Caribbean can hardly be distinguished from that of a hotel in Kuala Lumpur; Baltic super ferries now resemble American shopping malls, complete with McDonalds hamburger franchises. In the absence of genuinely original thought the same lexicon of space-filling decoration is constantly recycled with predictable monotony – marble, mirrors, brass and candelabra. In contrast, the quiet eloquence and immaculate detailing of the ships designed by Kay Fisker and Kay Kørbing lingers in the consciousness as a stinging rebuke to these later aberrations.

Bruce Peter MA RCA
The Glasgow School of Art
January 2004

A nostalgic view of the port of Esbjerg in 1966, taken from the after decks of the *Kronprins Frederik* and showing her sister, the *Kronprinsesse Ingrid*, tied up at the quay. (I.A. Glen)

The *Vistafjord* displays her graceful lines – developed by the Norwegian naval architect Kaare Haug – to good effect as she motors through the English Channel in this mid-1980s photograph. By this time, her funnel had been repainted in the attractive livery of Cunard Line. (FotoFlite)

DANISH SHIP DESIGN 1936-1991

INTRODUCTION

Kay Fisker (extreme right) enjoys dinner on the *Hammershus* with (right to left) his son, Steffen, wife Gudrun and Thorkil Lund, chairman of the '66' Company. (courtesy of Pauli Wulff of Poul Kjærgaard Architects)

Denmark is famed the world over for its modern movement architecture, furniture and product design, all of which have become renowned for their understated elegance, for their skilful manipulation of space and light, for their attention to exquisite detailing and for their respect for tradition and cultural continuity.

Such ingenuity was vital in a country built largely upon glacial outwash – a combination of sand, grit, chalk, clay and flint. Thus, vernacular building in Denmark

Kay Kørbing aboard the *Kong Olav V* in 1968, during the ship's maiden voyage for DFDS. (courtesy of Kay Kørbing)

used a limited palette of materials and the Danes were superb craftsmen. Clay could be made into bricks and tiles; roofs could be thatched with straw and heather and, above all, timber had many uses – from structural framing to small-scale ornamentation. 'Byggekunst' is a Danish word with no precise English equivalent, but which can loosely be translated as 'the art of building'. The inference is that the creation of order and beauty through the precise co-joining of architectural elements is in itself a great skill to which architects and designers should aspire.

Danish passenger liners emerged from this culture of craftsmanship and innovation. Already, by the 1920s Danish ships were technically advanced. For example, they were among the first to make widespread use of marine diesel engines. The Danish Burmeister and Wain (B&W) shipyard and engine works in Copenhagen developed the world's first ever marine diesel for an ocean-going ship when the passenger-cargo liner *Selandia* was delivered to the East Asiatic Company in 1912 (the first B&W marine diesel had been tested in 1898). Later in the 1920s, B&W developed an uprated diesel which, after further development, eventually became the most widely used marine power plant. Diesels were more economic than steam propulsion and took up less space, leaving more for profitable passengers or cargo. Besides, unlike Britain, which built coal-fired steamships until the Second World War, Denmark had no coalfields of its own and coal was bulky and expensive to import. Moreover, early marine diesel engines were notoriously noisy and often produced unpleasant vibrations, and so steam initially remained the more popular choice – at least in Britain.

The use of motor propulsion also had implications for the appearance of ships. The *Selandia* and her later sisters for the East Asiatic Company remarkably had no funnels and the exhaust instead went up pipes attached to one of their three masts. Motor passenger liners, on the other hand, tended to have shorter, wider funnels than steamers as these could also house generators, ventilators and other unsightly machinery to give these ships a more modern uncluttered profile.

Yet, until the late-1920s, ship interiors tended to reflect the style of stately homes and grand hotels as ship owners tended to be conservative. They believed that decorative interiors would impress passengers and might help them to forget that they were actually at sea. Traditionally, they selected a firm of naval architects to design their vessels and separate interior designers to decorate them. While the Germans and Italians in particular favoured voluptuous baroque for the first class saloons of their most prestigious liners, the Danish national flagship, the *Frederik VIII*, built in 1913 at Stettin for DFDS, had interiors by Carl Brummer in comparatively restrained neo-classical style with only occasional recourse to baroque flourishes. Born in 1864, Brummer was the first Danish architect to specialise in the problems of ship design, as well as architecture on *terra firma*.

At the turn of the century, he began working at Burmeister and Wain, firstly by enlarging the shipyard's estate of buildings and then also with the ships themselves. DFDS's sistership *Kong Haakon* and *Dronning Maud*, built in 1906, were the first. Most significantly, Brummer was the interior architect of the revolutionary motor ship *Selandia*, described above.

THE ORIGINS OF MODERNISM

All over Europe in the late 19th and early 20th centuries, a desire arose for a new form of architectural expression that was not dependent upon historical styles and which sought truth in terms of function and in the use of new forms of construction. Many of these experiments, which achieved fruition in the decade after the First World War, have been loosely defined as the Modern Movement, but have also been called The New Architecture or Modernism.

The pioneers of this new movement largely rejected applied ornamentation. Instead, they sought to embrace the machine age and the grace of utility apparent in iron, steel and industrially manufactured glass. The exploitation of such mechanically produced materials was crucial to the development of the Modern Movement. By the 1860s, iron was used structurally in many large buildings, but its unadorned expression in facades was almost unheard of. The realisation of its beauty, however, probably came about initially through the construction of iron bridges. The French architect and writer Viollet-Le-Duc (1814-79) [2] was probably the staunchest promoter of the expressive structural possibilities of iron. He called on architects to embrace manufacturing advances and not to hide architecture in the decorative languages of the past. In parallel, however, the search for clarity and coherent order once again looked towards classical rules of proportion as a means by which the constructive elements of modern building could be brought into harmony.

The aftermath of the First World War provided the nascent Modern Movement with renewed impetus and emphasis shifted to the promotion of its wider social and political agenda. Architects proposed to develop new structures to aid the social programmes required by the new order, in such a way as to provide spiritual uplift. In theory, new functional and egalitarian architecture would replace the feudal styles and distinctions of old, and perhaps help to create a cohesive new society.

One of the many reasons why the new generation of architects rejected the past was the belief that predetermined historical styles were in conflict with the uses of buildings. They believed that the function of a building as expressed through its plan should be the starting point of architecture. A secondary reason was the past's apparent failure to exploit materials to their full technical capacity, or to their full aesthetic value.

In contrast, modern architects and designers had little interest in superficial ornamentation. From the demand of rationalisation and machine-based mass production, it was a logical transition to advocating an aesthetic of smooth, uncluttered lines, stripped to their bare essentials. That theme was further inspired by the machines of travel – cars, aeroplanes and ocean liners. Aspiring modernist architects came to admire the external forms and masses of passenger ships as they embodied the combined virtues of the age, being utterly functional and shaped according to necessity, or so it was said. However, some were horrified by what they felt to be their excessively ornate interior decoration. In *Vers Une Architecture*, Le Corbusier used a Cunard Line publicity photograph showing its flagship, the *Aquitania*, to be longer and taller than all the most famous triumphal buildings in Paris. The young architect was clearly intrigued by the ability of the ocean liner to be at once technically functional, yet somehow awe-inspiring. He wrote about:

'...New architectural forms; elements both vast and intimate, but on man's scale; freedom from the 'styles' that stifle us; good contrast between solids and voids, powerful masses and slender masts...'[3]

Based upon a machine mythology, the aim of the modern movement was not just to imitate the technological process but to create a new species of modern objects, distinguished by their clean-cut complexion and rational sensibility. In architecture, one probable distinction of modern movement thinking, which, as we shall see, had consequences for ship design, was that a truly modernist building would be designed from the inside out with the volumes and fenestration of the exterior reflecting the internal organisation.

The turning point for shipboard interior design was the 1925 *Exposition Internationale des Arts Décoratifs et Industriels Modernes* in Paris, an exhibition of architecture, applied art and design for industry. Although the exhibition's title became associated with the 'Art Deco' style, the exhibition itself was stylistically eclectic, encompassing such avant garde exhibits as Le Corbusier's *Pavillon De L'Esprit Nouveau* and Melnikov's constructivist *Soviet Pavilion*. Kay Fisker (1893-1965) designed the Danish Pavilion. By the mid-1920s, Fisker was regarded as one of Denmark's most thoughtful and progressive architects. Yet, the pavilion was a tall, symmetrical composition in stripped-back neo-classical manner, somewhat resembling a mausoleum. From the outside, it gave little hint of Fisker's forthcoming modernity. Nonetheless, within, it housed an exhibition that was a brilliant showcase for the talents of many innovative Danish designers of the younger generation.

Only two years later, the Compagnie Général Transatlantique (known as the French Line) introduced its stunning *Ile De France* to the Le Havre – New York service. In terms of naval architecture, she was clearly a development from the earlier *Paris,* but was more remarkable on account of her highly innovative interior design, which mixed Art Deco motifs – exotic French Colonial, Egyptian, Mayan and 'jazz' imagery – with an entirely new and appropriate design language for ships – streamlining. Shortly thereafter, the first recognisably modern Scandinavian liner was the Swedish *Kungsholm* of 1928. A colourful, but refined essay in the modern romantic idiom, tinged with classicism, she was the work of Carl Bergsten, who had previously designed the Swedish exhibit at the *Exposition des Arts Décoratifs*. Yet, for all their modernity within, both the *Ile De France* and *Kungsholm* showed only a little external advancement over their immediate predecessors. For there to be a truly modern passenger ship, like-minded naval architects and interior architects would have to co-operate on an unprecedented scale to achieve the 'total form' required by modernism.

The Århus University campus, begun in 1933 and the result of an architectural competition held two years previously, perhaps best demonstrated Kay Fisker's approach to modern architecture. The various faculties were arranged amid parkland and make extensive use of yellow bricks, pan tiles and local timber. (Author)

THE FUNCTIONAL TRADITION

In Scandinavia, the Stockholm Exhibition of 1930 owed much in its architectural expression to ships. Designed by Eric Gunnar Asplund and other progressive Swedish architects, the exhibition received widespread and enthusiastic coverage in the European architectural press. The Swedes took to the new look very quickly, realising that simple white-rendered buildings looked very well in the bright Scandinavian sunlight. From then on, the modern movement dominated the Scandinavian architectural scene. Largely as a result of the Wall Street crash and the ensuing Great Depression, it was not until the mid-1930s that modernist principles in architecture and design could be applied extensively to a new generation of passenger ships (Denmark's merchant navy had been badly affected but, being largely an agrarian rather than an industrial nation, its economy overall was less seriously afflicted than those of Germany and Britain).

At that time, the most radical of the modernist theoreticians were arguing that architecture had to deal with such pressing social problems that there could be no room for historical references, hence the suggestion of a desire for functionalism. Kay Fisker corrected this revolutionary view with an evolutionary explanation, 'the functional tradition.' He found that the beauty of traditional building forms was achieved over many generations, eliminating what proved superfluous until what remained was the pure, utilitarian form, perfect in its simplicity. It was a kind of Darwinism applied to design, which closely mirrored the evolutionary approach used by naval architects, refining existing precedents and building on prior knowledge. Not surprisingly, Fisker's vision of modernity proved to be well suited to the design of passenger ships.

Kay Fisker's architecture developed from Denmark's building traditions – which emphasized a harmony of texture, supple articulation, primary three-dimensional geometric forms and Danish neo-classicism.

Danish neo-classicism had roots in Roman architecture - in Palladianism - and also in Nordic National Romanticism. While architects associated with early modernism in Germany and Austria, such as Adolf Loos and Peter Behrens, expressed admiration for the Prussian neo-classicist K. F. Schinkel, Danish architects pointed to their own great 19th century hero, C. F. Hansen, whose architecture was less bombastic and made subtle use of colours, materials, proportion and texture, as their favoured role model. Another important principle was an orientation towards simplicity, as evolved in Denmark's provincial building traditions, combined with a uniform and orderly co-joining of the architectural elements. Kay Fisker developed a similar attentiveness to these aspects of architecture and an equal aversion to what he viewed as frivolous ornamentation.

Neo-classicism remained as an orderly matrix in the minds of Danish architects who became interested in the modern movement during the 1930s. While International Modernism advocated a freer civilizing approach, linked to scientific analyses and the exploitation of new technology, Kay Fisker sought to reconcile history and civilization, the community and society, nature and culture. Indeed, to him, Denmark's traditional architecture of yellow bricks and tiled pitched roofs was nature transformed into culture, borne of a clear sense of the Danes' place in history and the world. Fisker designed on a readily accessible, human scale and with an egalitarian social outlook, emphasising deep-rootedness, significance, and sobriety. Perhaps Fisker's greatest skill lay in composition. He had an unwavering ability to create visually stimulating, yet harmonious facades simply through judiciously relating areas of unadorned brick and simple fenestration. As we shall see, his compositional dexterity, usage of traditional Danish building materials and felicitous attention to detail were invaluable foundations from which to contribute to the design of passenger ships and their interiors.

THE NEW SHIPS

By the 1930s, many Danes had more money to spend on holiday travel, and so the country's domestic short sea routes grew in popularity and there was a requirement for new passenger ships with some freight and car capacity. These were to be modestly sized comfortable vessels which relied not on unimaginable ostentation and statistical superlatives, but on the efficient use of space, economic propulsion and modern design.

Clearly, designing a ship involved co-ordinating a variety of skilled professionals, each with their own emphases and, sometimes, conflicting interests – naval architects, a variety of engineers, interior architects, financiers, the shipyard's directors and foremen and the ship owner. Shipbuilding was (and remains) an expensive process and the stakes were high for all involved, and so time was of the essence. This could make the role of architects, such as Kay Fisker and Kay Kørbing, challenging. According to Kørbing:

When working on the design of a building on terra firma, the architect has a far greater degree of control than when he is designing ship interiors. The end result of a ship's design, though, is deeply affected by how well the various constituents of the design team have co-operated and, I am afraid, it was sometimes a battle of wills to make the naval architects and engineers see things from my perspective. At the

The DFDS North Sea flagships of the 1960s are seen together at Esbjerg for the first time. While the 1964-built *England* manoeuvres off the berth, the brand new *Winston Churchill* is dressed overall and ready to load cars through her bow door. (Author's collection)

beginning of a new working relationship, it could sometimes be a struggle to persuade because the others were not necessarily trained to have the same type of co-ordinating design skills as architects and sometimes viewed me as an unnecessary imposition. The technical people in a shipyard didn't always see the "bigger picture' of how a completed ship would feel overall and they often had to be cajoled and persuaded to look beyond their own immediate contributions to the design process. Both Kay Fisker and I believed that ships should be 'total forms' – that is to say, their interiors should reflect their external appearances – and that their means of construction should be apparent both inside and out. My best projects were the ones where there was the closest co-operation between all of those involved and, when a completed ship was delivered, everybody involved could see why I had demanded certain things which they couldn't understand at the time.'[5]

Successful ship interiors had to solve conflicting requirements. On the one hand, they needed to reflect the technology and modernity of the vessels themselves – animated structures which creak, vibrate and have hissing ventilation systems. On the other, they were expected to combine an appropriate level of passenger comfort with a sense of security and solidity, while at the same time reflecting the glamour and adventure of a sea voyage. Perhaps the greatest achievement of Kay Fisker and Kay Kørbing's ship designs is that each ingeniously struck an appropriate balace between these demands. The use of beautiful natural materials – in particular dark wood panelling – and soft lighting created what the Danes call a 'hyggelig' atmosphere – warm, intimate, friendly and relaxed. With a drink and a good cigar to hand and a lively band playing, passengers could sit back in their comfortable armchairs as the ship swayed gently through the cold northern night. Many ephemeral qualities added to their unmistakable characters: the types of clothes and jewellery worn by passengers, the sounds of Danish, Norwegian or Swedish being spoken, the mixed aromas of cooking, perfume, French polish, tobacco smoke, diesel oil and the ping-pong of tannoy announcements echoing in dimly lit corridors. These helped to make modern Danish-designed passenger ships unique and special – from purposeful ferries at one end of the scale to the most elegant transatlantic liners and international cruise vessels at the other.

At Riva Trigoso in Italy in the Spring of 1968, the *Kong Olav V* nears completion while the *Prinsesse Margrethe*, to the right, remains shrouded in scaffolding. (Author's collection)

THE WORK OF KAY FISKER AND KAY KØRBING

CHAPTER 1
SHIPS DESIGNED BY KAY FISKER
MS *HAMMERSHUS*

The *Hammershus* is seen tied up at the Burmeister & Wain Shipyard in Copenhagen, shortly before delivery. (Author's collection)

In 1934, the Dampskibs-Selskabet af 1866 paa Bornholm (popularly known as the '66' Company), which traded between Copenhagen and the island of Bornholm in the Southern Baltic, ordered a new passenger ship from the Burmeister and Wain Shipyard in Copenhagen. The previous year, Thorkil Lund, an enterprising Bornholm businessman, had joined the firm's board of directors and set about steering the company in a forward-looking direction. Dissatisfied with the shipyard's own proposals, the '66' Company, reputedly at Lund's suggestion, requested Kay Fisker's assistance in the summer of 1935. Fisker was well-known on Bornholm as he had previously designed stations and other buildings for the Gudhjembanen railway, opened in 1916, and for the Handelsbanken in Rønne, completed in 1921. That summer, the youthful Poul Kjaergaard became an assistant in his Copenhagen office: 'A few days after accepting the commission, one early July day, a new, slightly naïve assistant walked into the drawing office on the top floor of the Lagkagehuset on Vodroffsvej. He was welcomed by Fisker with the question 'can you draw ships?' and with the then slim chances of finding work, his response had to be 'sure – why not?' The answer was accepted and he was allowed to start on Fisker's first ship design project.
'The drawing office was by present day standards rather small with only a half-dozen employees to work on the various projects… Over the whole hovered Fisker, at his prime in his early-forties, very active in the architecture school and the professional debates, but also with thorough daily supervision of all the work in his office…'[6]
At that time, Fisker's designing partner, C.F. Møller, was based in Århus, where he was supervising the ongoing Århus University project, leaving Fisker himself in

The *Hammershus* shows her crisp, modern lines and all-white livery to good effect in this view taken in Copenhagen. (Author's Collection)

control of the Copenhagen office. Prior to Kjærgaard's appointment, another assistant, Erling Frederiksen, had already made a series of preliminary proposals for the new Bornholm ship. Thereafter, Kjærgaard and Frederiksen were the job architects who refined and drew up Fisker's subsequent ship interior designs. (As the volume of this work increased, a third architect, Jørgen Grønborg, was appointed.)

As Fisker had been appointed late in the design process, time was unusually tight and the keel for the new ship was shortly due to be laid on the slipway. There were frequent intensive meetings with Thorkil Lund and the shipyard. Early on in the process, it became clear that strict limitations had been placed on Fisker's creative freedom as B&W had already agreed a tightly budgeted contract price. Besides, the yard had its own drawing office, which also prepared interior designs. So had Fisker's proposal been too outlandish or expensive, the project could easily have been given back to B&W. Moreover, Fisker and his assistants were on a steep learning curve. They had quickly to master naval architectural drawing standards, which used British imperial measurements and scales such as 1:96 and 1:24. The sheer and camber of a ship's hull had to be taken into account, as did the need to be much more precise about measurements than was then typical in the building industry, in which slight adjustments could easily be made on site.

'After the staff in the drawing office made the first investigations and sketches, Fisker early on determined the principles that he would follow in the work to come. Naturally, an evaluation of the traditional ship arrangement techniques of the time fed into this thinking. Fisker himself, through his constantly updated general knowledge, was well-versed in the latest developments, especially on the Atlantic run, where the ocean liners of the dominant countries fought to gain the leading position in terms of speed and internal grandeur. Of the Danish efforts, he suggested that the research should begin through familiarisation with the ideas previously laid down by the architect Carl Brummer, which were still common practice in Danish shipyards. Fisker had previously experienced this style at first hand on a voyage to the Far East, courtesy of H. N. Andersen, the powerful chief of the Danish East Asiatic Company.'[7]

The starting point for the new interior design was a set of drawings and renderings prepared by the shipyard and in the then-conventional neo-classical manner of Brummer's shipboard work. Fisker, however, had his own strongly-held views about how the new ship should look.

'Early on, Fisker had decided on his intentions and had secured the approval and support from the shipowner, which would give him greater influence over the shipyard in decision-making. He thought that the saloons on the promenade deck should comprise a complete spatial entity, rather than having a series of rooms of different characters, as was then the norm. Openness and visual unity would also ease the passengers' orientation. The staircases, smoking saloons and dining rooms, therefore, would have their walls and ceilings executed in a related range of materials, treated and assembled in a similar fashion throughout.

'In first class, smooth natural wood panels were assembled and fixed with brushed brass strips, which were screwed in place. From the substantial selection of woods available at the shipyard, a slim elm veneer was selected to cover the walls running the length of the ship and the ceilings, while palisander veneer was chosen for the transverse bulkheads and partitions. This was a suitable background for the marquetry decorations, which had been specially created by craftsmen. For the more modest third class accommodation, smooth, white-painted panel walls and ceilings were chosen. In the third class smoking saloon, the upper walls were decorated with a number of aerial photographs of settlements and landscapes in Bornholm.'[8]

In designing the interior of the new ship, Fisker closely followed the rational principles which governed all of his architectural and design endeavours – to use traditional, hard-wearing materials which had been shown to stand the test of time, in a stripped back, unadorned manner, emphasising their intrinsic beauty. The detailed outfitting of the ship required particularly careful thought for it was intended to sail in all weathers on what was a lifeline service. The furniture and fixtures would, therefore, have to withstand punishing usage in rough winter weather when some passengers might even be seasick. While traditional shipboard furniture was understandably solid-looking, such modern-looking furniture, as was then available, was not intended for such treatment, and accordingly most had to be purpose-designed by Fisker and his colleagues. Ashore, it was then accepted practice that good architects would design the furnishings for their buildings to achieve visual harmony and coherence.

'As many fittings as possible had to be fixed in position and individual items of furniture needed to have a certain amount of substance to stay in place in heavy weather. These requirements were fulfilled through the use of built-in sofas, tables and substantial leather-clad chairs. For Thorkil Lund, the arrangement and equipment of the cabins was also an important area of design for Fisker to bring up-to-date. In these spaces, walls and ceilings were carried out with smooth white-painted surfaces. All metalwork – the berth frames, plumbing and ironmongery – was chromed. Fabrics for the curtains and carpets were specially made from drawings or selected from the leading weavers and fabric printers of the time. The owner's suite was given a superior quality of fittings.'

The motor ship *Hammershus* was delivered to the '66' Company in June 1936 and, although very small at only 1,726 tons, she provoked much interest on account of her strikingly modern design. While the architecture and design press enthused over her modernity, she was hailed in the Copenhagen newspapers as 'en helt lille *Queen Mary*' (a complete little *Queen Mary*) – a compliment indeed as the new Cunard liner was then the world's largest, fastest and one of the most glamorous ocean liners.[9] Although the two ships could not have been more different in terms of size, the comparison was perhaps appropriate. As with the famous Cunarder, the new *Hammershus* had a cruiser stern and purposeful, modern lines. Most importantly for the *Hammershus*'s proud owners, she proved a highly popular ship with passengers.

In April 1940, Denmark was invaded by the Germans and on 4th August disaster struck as the *Hammershus* hit a mine shortly after sailing from

The first class entrance hall of the *Hammershus* was finished in a variety of wood veneers with brushed chromed balustrades and matching joints between the timber panels. (courtesy of Pauli Wulff of Poul Kjærgaard Architects)

In the dining saloon, the outboard walls were clad in darker veneers than the bulkheads and ceiling. Note the buttoned leather furniture, designed specially for the ship, and the recesses around the windows, containing roller blinds. Such careful attention to design detail was the essence of Fisker's approach. (courtesy of Pauli Wulff of Poul Kjærgaard Architects)

Copenhagen. Fortunately, the SS *Benedikt* rescued the passengers and the vessel was towed onto a sandbank. Repairs cost one million kroner and were paid for by war insurance. In January 1944 German marines seized the *Hammershus*, but the '66' Company recovered her in May 1945 and loaned her to the British forces for trooping duties. The ship finally returned to the Copenhagen-Rønne service in 1947, but only after a thorough refurbishment. She was sold to the Danish Navy in 1964 for use as a depot ship named *Henrik Gerner*. Decommissioned in 1975, she was scrapped the following year in Odense.

The first class saloon was a rectangular space occupying the full width of the forward superstructure. On the rear bulkhead, there was a marquetry panel showning an outline of Bornholm and its major settlements and attractions. Note that much of the furniture is built-in and all the movable items are very solidly designed as the *Hammershus*, being a mail boat, was required to sail in all but the very worst weather conditions. (courtesy of Pauli Wulff of Poul Kjærgaard Architects)

DANISH SHIP DESIGN 1936-1991

MS *KRONPRINS OLAV*

Adjacent to Kvæsthusbroen in Central Copenhagen stood the handsome headquarters of Det Forenede Dampskibs Selskab (The United Steamship Company) – one of Denmark's most prestigious shipping lines, whose services then reached all over the western hemisphere. The new *Hammershus* must have provoked much curiosity amongst DFDS's senior management – especially the firm's forward looking administration director, J.A. Kørbing.

The company had recently ordered a new ship for its premier Copenhagen-Oslo service from the Helsingør Skibsværft, which was also a DFDS subsidiary. Thus, a few months after the *Hammershus* entered service, Kay Fisker was invited to meet Kørbing as he wanted Fisker's assistance to design the new ship's passenger accommodation.
'Many of these frequently lengthy meetings had to be carried out at the shipyard in Helsingør and clearly demanded Fisker's personal attendance. The trips there in the yellow Opel Super Six were longer and more difficult for Fisker to handle, but from the point of view of his assistants, they became memorable. Once on the road, Fisker was quite willing to make detours to examine and criticise buildings and landscapes, which he thought worth investigating.'[10]

A great advantage to Fisker was the fact that Kørbing had involved him much earlier on in the design process than in the *Hammershus* project and so he had the possibility of making a greater input to the new ship's overall layout. Better still, he would now be working in close collaboration with the highly regarded young naval architect, Knud E. Hansen, who was then at Helsingør Skibsværft. Ever since the *Hammershus* had entered service, Fisker and Hansen had corresponded to learn more from each other about innovations in naval architecture and interior design.

A pre-war view of the *Kronprins Olav* shortly after delivery to DFDS in 1937. In terms of design, she represented a fusion of ideas from the Hammershus with new innovations jointly devised by Kay Fisker and Knud E. Hansen. Being a larger ship, she had more elongated, graceful lines, accentuated by the horizontal contrast of the black hull and white superstructure and the curvaceous form of the first class saloon, located forward. (Author's collection)

Born in 1900 in Helsingør, Knud E. Hansen had studied naval architecture at Danmarks Polyteknisk Læreanstalt. Following his graduation in 1925, he gained practical experience in shipbuilding at a number of yards, both in Denmark and abroad. Having joined the design department of Helsinør Skibsværft, the first passenger ship he worked on was the Bergen Line's MS *Venus*, completed in 1931. Her crisp, modern lines and rational planning certainly influenced his own thinking, which was very much in tune with Fisker's own architectural philosophy. Working together, the two men were to produce for DFDS one of the most significant passenger ships of the 1930s and Fisker was closely involved in both the external and interior design:

'DFDS wanted the *Kronprins Olav* to perpetuate the shipowner's earlier tradition by having the hull painted black. The clear demarcation between the black hull and white superstructure was therefore important so that, together with the slant of the masts and funnel, the elegant lines of the ship would be emphasised. The shipowner and the shipbuilder supported the proposals made by Fisker to achieve this. The shape of the front of the superstructure was very significant for the perception of the ship as a harmonious entity. In the *Kronprins Olav*, this was given a semi-circular treatment as opposed to the rectilinear saloon designs of earlier ships.'[11]

Thus, the *Kronprins Olav* was sleek and streamlined with a high bow profile, a cruiser stern and a long, low superstructure with a tapering funnel. New advances in steel technology enabled the vessel to have a much more curvaceous silhouette than had hitherto been possible. Thanks largely to Fisker and Hansen, ship superstructures in future would be swept back with complex double curvatures and, later, domed funnels to match. Consequently, their designs would be better integrated and their forms more harmonious.

The striking modernity of the first class saloon is emphasised by the perimeter lighting and the recessed ceiling in the middle of the space. The lack of cluttering ornamentation makes it appear to be far larger than it actually was. (courtesy of Pauli Wulff of Poul Kjærgaard Architects)

In the interior, Fisker developed the design principles he introduced in the *Hammershus*. The layout and shapes of the spaces were finalised early on in the design process, after which as much time and attention as possible was spent on detailing the interiors. With their knowledge and experience gained from the *Hammershus* project and the more generous timescale, it was possible for Fisker and his assistants to work closely with the shipyard to achieve a very high standard of finish.

The walls in all the first class saloons were lined with matt-varnished sycamore with ceilings in similarly-finished hazel. Joints between the panels were matt-chromed metal. Throughout, the lighting was indirect, either concealed behind

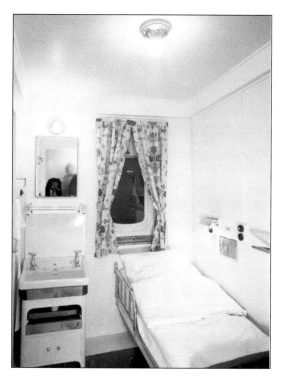

While the *Kronprins Olav*'s public rooms were very modern, unfortunately the cabins were less so as DFDS insisted on using its existing standard design, as shown here. (courtesy of Pauli Wulff of Poul Kjærgaard Architects)

coves or around the perimeter of floating suspended ceilings. Fisker designed a new range of furniture for the *Kronprins Olav* – high-backed armchairs with winged headrests and long curved banquettes. In the semi-circular first class smoking room, these were in red leather, which contrasted with the black linoleum flooring used throughout and also referred to the DFDS livery. In the adjacent first class hallway, the *Kronprins Olav* had a grand staircase with a portrait of the Norwegian crown prince strategically placed on the bulkhead behind. Fisker however had only marginal involvement in the design of the cabins, which were to DFDS's own standard specifications.

A group of high-backed armchairs designed by Fisker in the saloon. (courtesy of Pauli Wulff of Poul Kjærgaard Architects)

Work progressed rapidly at the shipyard. The hull was launched in an advanced state of completion in June 1937 and on 18th December, the ship left Copenhagen on her maiden voyage to Oslo. With almost everything specially designed – from the overall profile down to clocks and ashtrays – the *Kronprins Olav* was a wonderful advertisement for modern Danish design and one which was to have

The first class dining saloon was a uncluttered rectangular space with long wings containing more seating on either beam. This layout was repeated on subsequent ships designed by Fisker for DFDS. (courtesy of Pauli Wulff of Poul Kjærgaard Architects)

a great influence on subsequent Scandinavian passenger vessels. Because of their lightness and lack of cluttering decoration, her interiors gave the impression of being far more spacious than they actually were and once in service, she was immediately popular both with passengers and the architectural press. Fisker was understandably proud of the completed ship, so much so that he personally wrote an article on her design for a DFDS's press release in December 1937:

'In shaping the saloons, an attempt has been made to deviate from the tradition whereby saloons were clearly separated by a hallway and in which each saloon was given an individual treatment of walls, ceilings and floors. In the *Kronprins Olav*, by contrast, we tried to to draw the hallways in as links between the saloons so that

The *Kronprins Olav*'s entrance hall must have been particularly effective as it combined light wood veneers, brushed aluminium balustrades, black linoleum flooring and red leather-clad seating – reflecting both the DFDS livery and the Danish national colours. (courtesy of Pauli Wulff of Poul Kjærgaard Architects)

The second class saloon on the *Kronprins Olav* featured built-in furniture in a white-painted space. (courtesy of Pauli Wulff of Poul Kjærgaard Architects)

the whole public room arrangement [in each class] appears as an entity. Wall and ceiling materials are identical. The mode of illumination is the same throughout and wide glass doors permit an unobscured view through the full length of the saloons. Altogether, these tactics assist in making the saloons spacious, warm and friendly and also improve passenger orientation. Consequently, each saloon derives its atmosphere from its shape and from the shapes and colours of the individual pieces of furniture.'[12]. At the 1938 Spring Exhibition at Charlottenborg, Fisker, by this time also the Professor of Architecture there, mounted a display of photographs and architectural drawings of his firm's ship design schemes. By then, more were on the way.

The *Kronprins Olav* was sold by DFDS in 1967 to Italian owners for use as a ferry to Corsica and, later, across the Bay of Naples. After rebuilding, during which the cabin areas were dismantled to create a stern and side-loading car deck, the ship re-entered service as the *Corsica Express* of Corsica Line. Later known as the *Express Ferry Angelina Lauro* and the *Capo Falconara* for successive Italian owners, she was only sold for scrap in December 1986 after a career of nearly fifty years.

The *Corsica Express* arrives in Genoa in 1968. This photograph was taken by Kay Kørbing when he was in Italy working on the five ships built there for DFDS. (Kay Kørbing)

1939 MS *HANS BROGE* AND REBUILDING OF MS *C.F. TIETGEN* (built 1928)

The *C.F. Tietgen* as originally designed in 1928. (Author's collection)

DFDS next turned its attention to the modernisation of its busy Copenhagen-Århus domestic overnight service, which linked two of Denmark's principal cities. A new ship, to be named *Hans Broge*, was ordered from the Helsingør Skibsværft and, at the same time, it was decided to give the existing ship, *C.F. Tietgen* (built in 1928), a major rebuild. DFDS was anxious that the interior design of these ships should be as close as possible to the high standard of the *Kronprins Olav* and so, naturally, Fisker was again chosen as architect.

'These ships gave Fisker the chance to concentrate his attention on a number of distinct problems which came to light from the two earlier projects and to try out some other design concepts and details. With the benefit of experience from the previous ships and with the already very good working relationships with the shipowners and shipyard, the work with these new projects made steady and easy progress. This was vital as their construction proceeded in record time. Only six months passed from the laying of the *Hans Broge*'s keel to delivery. At the same time the shipyard itself was being modernised and the 10-year old *C.F. Tietgen* was being rebuilt and enlarged.'[13]

The design of the new *Hans Broge* was less radical than that of the much-acclaimed *Kronprins Olav*. The explanation for this was that, for operational reasons, DFDS had decided that she and the rebuilt *C.F. Tietgen* should be as close to sister ships as possible, a decision which seriously inhibited the design possibilities. (The *C.F. Tietgen* was designed and decorated in the late neo-classical style, the standard for Danish passenger ships set in the pre-First World War era by Carl Brummer.) None-the-less Fisker and his colleagues were determined to make the best of a difficult situation:

DANISH SHIP DESIGN 1936-1991

The *C.F. Tietgen* after rebuilding. The wheelhouse and forward superstructure is almost the only recognisable feature of the original design. (Author's collection)

'The public circulation spaces and saloons were here, as always, an important area. At that time, DFDS ships had two classes (this continued until 1970). In the new Århus vessels, the difference in outfitting between the classes was markedly reduced. This was achieved by substantially improving on second class.'[14]

The *Hans Broge* and the rebuilt *C.F. Tietgen* were completed by June 1939 and quickly introduced on the Copenhagen-Århus route, with nightly sailings in each direction. Both externally and within, they represented a compromise between the *C.F.Tietgen*'s original 1920s design and Fisker's modernity. The interiors, though, were similar in detail to those of *Kronprins Olav* – but unfortunately lacked any of the

The first class smoking saloon of the *C.F. Tietgen*, as originally designed in Carl Brummer's stripped back neo classical manner. The design of this saloon was typical of Danish passenger ships of all sizes from the turn of the century until the mid-1930s. (courtesy of Pauli Wulff of Poul Kjærgaard Architects)

Effectively the same space on the *Hans Broge*, as designed by Kay Fisker. The clean lines, concealed lighting and modern furniture must have been something of a culture shock for passengers, conditioned to expect rather a more traditional approach from DFDS's Århus ships. (courtesy of Pauli Wulff of Poul Kjærgaard Architects)

latter's elegant curving forms. The Copenhagen-Århus service was shortly halted by the Second World War – by German invasion and by fuel shortages.

In 1944, both ships were commandeered by the Germans and crudely rebuilt as troop transports for the German navy. DFDS retrieved them in poor condition from Flensburg and Sassitz after the German capitulation and, after being used by the allies as refugee ships, they were refurbished and returned to the Copenhagen-Århus route. The two were further rebuilt and lengthened in 1954 and continued in service until 1969. The elderly and much-altered *C.F. Tietgen* went straight to the shipbreakers, but the *Hans Broge* was sold to the Greek Epirotiki Line and named *Achilleus*. Intended for cruises in the Cyclades, this vessel was instead chartered to the Ånedin Linjen of Mariehamn for short Baltic cruises from Stockholm to the Åland Islands, finally going for scrap in 1983.

The dining saloon of the *Hans Broge* is similar in design, but features shutters to close over its windows, rather than the more conventional blinds or curtains. Fisker reputedly did not like curtains as he felt that they made rooms look less tidy. (courtesy of Pauli Wulff of Poul Kjærgaard Architects)

DANISH SHIP DESIGN 1936-1991

1940 MS ROTNA

The recently-introduced *Rotna* is seen at the beginning of the Second World War wearing Danish neutrality markings. (Author's collection)

Also in 1938, the Dampskibs-Selskabet af 1866 paa Bornholm required Fisker's assistance to help with the design of an improved sister to the much-praised *Hammershus*. The new ship was named *Rotna*, the Latin name for Rønne, Bornholm's main town, at the planning stage. Although the ship was very similar in external appearance and general arrangement to the *Hammershus*, Fisker used the benefit of experience gained in the interval to improve aspects of his interior design strategy. One obvious innovation was the ceiling in the first class hall and dining saloon. In order to maximise the feeling of spaciousness in areas with fairly low deck heights, he devised a remarkable wavy design which bowed upwards between the frames and ventilation channels (incidentally, this was similar to the ceiling in the canteen of his university buildings in Århus). As the windows were evenly spaced between the frames, the result looked refined and appropriately nautical. Unfortunately, 'these ceilings proved to be contrary to the well-being of the passengers as in rough weather there appeared to be waves both inside and out, so after a couple of years, it proved necessary to get flat ceilings installed.'[15] The first class saloon was rectangular, with none of the streamlining of the *Kronprins Olav* – either to make more efficient use of space in what was a very compact general arrangement, or because the B&W drawing office was less avant garde in its approach to styling than that of Helsingør Skibsværft. Even so, Fisker devised a most intriguing lighting scheme which took the form of upturned copper 'mushrooms' to cast light upon the elm veneered ceiling panels, which then reflected a warm tone

The first class hallway and dining saloon initially featured remarkably wavy plywood ceilings to maximise the height between the ship's horizontal structural members. While the effect was certainly striking, these features were short-lived and were replaced with a more conventional design during an early refit. (courtesy of Pauli Wulff)

The first class saloon featured upturned 'mushroom' light fittings so that the wooden ceiling finish would reflect a warm glow into the space. On the rear wall is a marquetry panel of Bornholm. The seats are upholstered in cream buttoned leather. (courtesy of Pauli Wulff of Poul Kjærgaard Architects)

A fine image of the *Rotna* in The Sound, showing the smart '66' Company livery to good effect. The funnel mark is white rectangle featuring a postal horn with a crown – a reminder of the vital 'lifeline' the company's ships provided to Bornholm for over a century. (Author's collection)

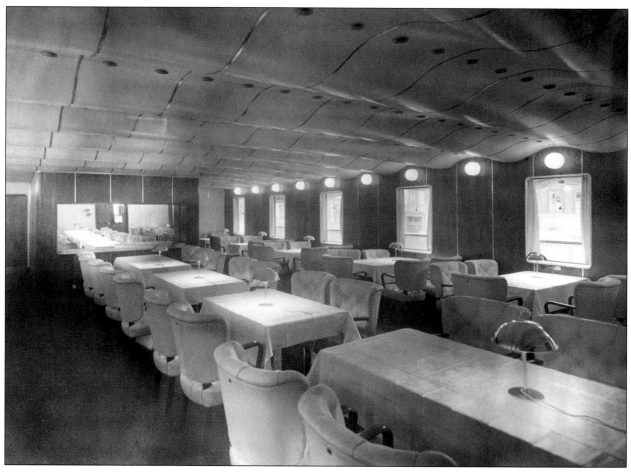

As with the hallway, the dining saloon initially had an undulating plywood ceiling. (courtesy of Pauli Wulff of Poul Kjærgaard Architects)

After rebuilding with a flat ceiling, the dining room is pictured again toward's the end of the Rotna's career in the mid-1960s. (Bornholmstrafikken)

back into the space. Direct lighting was provided by small table lamps. The *Rotna* was introduced in 1940, but was laid-up shortly thereafter, another victim of the war-time fuel shortages.

The *Rotna* spent much of the war laid up in Rønne harbour and was slightly damaged by Russian attacks during the liberation of Denmark. After repairs at Burmeister and Wain in Copenhagen, she was returned to service and continued until 1969. For a few months in the spring of 1965 she was renamed *Hammershus*. Sold to a German firm in December 1969, she was resold to Spanish owners for use as a floating hotel ship in Barcelona. Unfortunately, while being towed to Barcelona, the former *Rotna* foundered in the Bay of Biscay.

MS *KRONPRINS FREDERIK* AND MS *KRONPRINSESSE INGRID*

With her external design obviously developed from that of the *Kronprins Olav*, the new *Kronprins Frederik* made a bold, modernistic impression when finally introduced on the prestigious Harwich-Esbjerg route after the Second World War. (Author's collection)

Following the great popularity of the *Kronprins Olav* and the improved Århus ships, in the winter of 1938-9 DFDS turned its attention to providing better ships for its important Esbjerg-Harwich route to cope with the growing tourist and business traffic between Denmark and Britain. The existing quartet of ships on the route were essentially cargo vessels with some passenger space. DFDS proposed to build two larger ships in quick succession. To familiarise themselves with what the competition was offering, Fisker was invited to accompany the DFDS directors Garde and Quist on an 8-day fact-finding tour, sailing on a variety of recently introduced overnight ferries to acquaint themselves with contemporary design practice in other countries. Because of other professional commitments, Fisker had to withdraw from the trip, but he sent his young assistant Poul Kjærgaard instead:

'Swedish Lloyd's SS *Suecia* sailed from Gothenburg to London. This was a ten-year old steamship built in England and in its interior design, an exemplar of the English tradition with each saloon having its own style, as in a grand English manor house. Next, a visit was made in Liverpool to a completely new ship, MS *Munster*, which was on the Liverpool-Dublin route. In contrast to the stately Swedish ship, here there was thoughtless and excessive use of the latest inventions. For example, the ceiling illumination in the saloons consisted of coils of garish neon tubes. Following on, we sampled the Bergen Line's new Italian-built motorship *Vega*, operating between Newcastle and Bergen. The verdict of the DFDS management on the internal arrangement of the saloons was 'varied but modern and a little pretentious as in the Italian transAtlantic liners.' In Norway, we looked at another newbuild, MS *Black Watch*, which also sailed to England and which in size and outfitting was much like MS *Vega*. Note was made of the generous but eclectic presentation of Norwegian applied art. A substantial report was completed on return [to Copenhagen] with the *Kronprins Olav*. It concluded that while it was possible to learn from individual details in the ships surveyed, as far as the overall concept was concerned, everything indicated that in the new England boats, DFDS should build on its own standards as most recently demonstrated in the *Kronprins Olav*. This was a decision which Fisker was the first to approve of.'16

An order was placed with Helsingør Skibsværft in the spring of 1939 for the first of the new sister ships and, according to Poul Kjærgaard, work progressed well:

'The working atmosphere was the best possible. There were close relationships between the shipowner and the shipyard and Fisker and his drawing office gave them both excellent cooperation. It was possible, therefore, to carefully consider problems at an early stage and to get time to exchange views without being forced into hasty decisions. From the viewpoint of the interior architects, it was particularly important to have such a relationship with the engineers at the shipyard. In the upper decks, there was potential to dissociate the locations of the partitions between the saloons from the configuration of the main structural framing, which followed right through the whole ship from keel to top. This created great freedom to form and furnish the saloons in relation to window openings and other relevant features. It also became possible to increase the ceiling heights substantially by fitting the ventilation ducts in a new way within the wall constructions. All these

Among the most outstanding features of the *Kronprins Frederik* was the first class hallway, which featured a highly curvaceous stairway with a portrait of the much-admired Danish crown prince hanging behind. (courtesy of Pauli Wulff of Poul Kjærgaard Architects)

The first class saloon was semi-circular and featured high-backed black leather settees with winged headrests. As with all of the ship's original interiors, this space and all of its contents were destroyed when she caught fire at Harwich. One setee – probably a spare – was recently auctioned in Copenhagen, however, for over 20,000 Danish kroner (around £2000). It was probably a unique survivor because Kay Fisker designed furniture specially for each one of his projects. (courtesy of Pauli Wulff of Poul Kjærgaard Architects)

The first class dining saloon extended along either beam amidships in a U-shape around the galley so that the majority of passengers could enjoy a sea view. (courtesy of Pauli Wulff of Poul Kjærgaard Architects)

developments were of considerable importance to the spatial effect of the saloons. 'In lining the walls with panels, a new method of assembly was found, using fixing points in the panels instead of screwed on metal strips as these had exhibited maintenance problems. Smaller saloons received special wall coverings. In the bar, light-coloured leather was used, while the writing room featured braided Japanese mats. To illuminate the smoking saloon, Poul Henningsen's half-globe PH lights were built into the ceiling. For this particular purpose, Henningsen specially designed a bigger and flatter globe.'[17]

Before construction of the superb new 3,895 ton *Kronprins Frederik* could be commenced, the war broke out in September 1939. At first, work proceeded as planned and the keel was laid that November; when Denmark was invaded in April 1940, the ship was already well advanced. After a short break due to materials shortages, the hull was launched in June 1940. A year later, the *Kronprins Frederik* was delivered to DFDS. By this time, any possibility of using the ship for its intended purpose had vanished. Prudently, DFDS removed the furniture and other moveable fittings for safekeeping. The propellers and vital pieces of technical equipment were also detached and hidden to prevent the ship being used by the Germans. With the hull camouflaged, she was laid up in Copenhagen's South Harbour basin next to the royal yacht, *Dannebrog*. DFDS's decision to render its new flagship unserviceable later proved to be a good one, for in November 1944, the Germans decided to commandeer a large number of usable Danish ships for their own military purposes – but not the incomplete *Kronprins Frederik*.

After Denmark's liberation, the ship was towed back to Helsingør for completion in November 1945. Four months later, the *Kronprins Frederik* was finally ready to enter service – five years later than DFDS had intended. In appearance, the ship was very much an enlarged version of the *Kronprins Olav* and looked particularly handsome in DFDS's livery of a pale grey hull with red boot topping and buff masts. The low streamlined funnel was black with a broad red band, but as the base was hidden behind lifeboats, it appeared to be red with a black top.

Even the *Kronprins Frederik*'s second class accommodations were designed to a high standard. The smoking room has an appropriately masculine atmosphere with wood panelling and leather-clad furniture. (courtesy of Pauli Wulff of Poul Kjærgaard Architects)

Within, the ship had very high quality accommodation for only 358 passengers spread between first and second class. The elegant semi-circular first class smoking room filled the forward part of the saloon deck and to the rear there was a small curving cocktail bar. The hallway had a grand double staircase and immediately aft was the first class dining saloon, panelled in teak. Second class public rooms were located towards the stern. The ship was a triumph of miniaturisation and no space

The *Kronprinsesse Ingrid* is seen early in her career approaching the berth at Harwich Parkeston Quay. (author's collection)

was wasted. The stairways were steep and the corridors narrow, but cabins were squeezed into all kinds of unlikely corners even around the engine room and cargo holds. In contrast, there were a couple of luxury state rooms with separate sleeping and sitting areas.

The *Kronprins Frederik* was such a success that an identical sister ship, to be named *Kronprinsesse Ingrid,* was quickly ordered in 1947. Meanwhile, the war-damaged *Kronprins Olav*, which had been seized by the Germans in 1944, and was found in a sorry state in Rendsburg after the German capitulation, was also towed to Helsingør for refurbishment in the autumn of 1945. As it was vital that this vessel should also be made serviceable for use as a British troopship, this work was not as thorough as DFDS would have liked, so when finally returned to her owners in March 1946, she was given a second, much more substantial rebuilding emerging as a practically new ship. Kay Fisker had designed entirely new interiors in a manner similar to those in *Kronprins Frederik* and the forthcoming *Kronprinsesse Ingrid.*

The *Kronprinsesse Ingrid*'s first class interiors were rather more luxuriously appointed than those of the older *Kronprins Frederik*. Here we see the first class, writing room. (courtesy of Pauli Wulff of Poul Kjærgaard Architects)

The *Kronprinsesse Ingrid* entered service in June 1949, just in time for the busy summer tourist season. When at Harwich Parkeston Quay on alternate days, the two ships provided a stark contrast to the Hook of Holland-bound railway steamers. Even the newest of these seemed a little old fashioned in comparison with the sleek and brightly coloured Danes.

It was while tied up at Parkeston Quay on 19th April 1953 that the *Kronprins Frederik* suffered a destructive fire, which began in her engine room. She was badly damaged and the local fire brigade pumped in so much water that the next day she keeled over at the berth. It was four months until the wreck could be re-floated and it appeared that she would be a write-off. Instead the *Kronprins Frederik* was towed to Helsingør for another thorough rebuild, during which Kay Fisker had an opportunity to redesign the interiors as well. His revised scheme was much more luxurious than its predecessor with sumptuous blue and cream upholstery in the first class smoking room, complementing the stunning cocktail bar, adjacent, in which the bar front, walls, ceiling and furniture were clad in brown padded leather.

The *Kronprinsesse Ingrid* was sold to the Rederi og Handelsselskabet Montana in 1969 to be used on tax-free shopping day cruises from Copenhagen to Halmstad in

DANISH SHIP DESIGN 1936-1991

The fire-ravaged *Kronprins Frederik* lies capsized at Harwich Parkeston Quay as the venerable former Vlissingen steamer *Mecklenburg* of the Dutch SMZ lies off shore and British Railways (Eastern Region's) more modern *Arnhem* waits at the quay ahead. (Author's collection)

Once she was raised on an even keel, it became clear how badly damaged the *Kronprins Frederik* was. Clearly, a major reconstruction was required. (Author's collection)

The new interiors were similar to those of the *Kronprinsesse Ingrid*. In this DFDS publicity photograph, we see the first class dining saloon, finished in rich wood veneer and with lighting by Poul Henningsen recessed into the ceiling. (Author's collection)

Sweden as *Copenhagen,* but the plan was not a success and, after only three months, the ship was laid up and offered for sale. She was quickly sold to a Greek shipowner, Costas Spyrou Latsis, and entered Mediterranean service as the *Mimika L*, sailing to the Cyclades from Piraeus. From 1978, she was owned by Gerasimos Ventouris and re-named the *Alkyon*.[18] Laid up in 1983, she was sold for scrap in Pakistan two years later.

The *Kronprins Frederik* continued in the DFDS fleet until August 1974 when this stalwart was laid up at Esbjerg. Two years later, she was sold to Egyptian owners for use as a pilgrim ship in the Red Sea. It was only to be a brief new career for in December 1976 after only eight months in service and carrying the name *Patra*, she caught fire and sank, killing 102 passengers.

DANISH SHIP DESIGN 1936-1991

MS KONGEDYBET

The *Kongedybet* displays her purposeful, modern lines in a series of images of the ship shortly after entering service for the '66' Company between Copenhagen and Rønne. (courtesy of Pauli Wulff of Poul Kjærgaard Architects)

After Denmark's liberation, the Dampskibs-Selskabet af 1866 paa Bornholm again sought Fisker's assistance, first to advise on the rebuilding and modernisation of the war-damaged *Hammershus* and *Rotna*. Each was repaired by Burmeister and Wain in 1945-6 and returned to the Copenhagen-Rønne route. The '66' Company soon experienced a significant upturn in passenger numbers — especially summer tourists — and so in 1949 Thorkil Lund again asked Fisker to help design a third new ship of a similar design to the existing duo to finally to make good war losses.[19] In 1950, an order was placed with B&W and the company announced that the new ship would be named *Kongedybet*. It had been ten years since the *Rotna* had been designed, and so Fisker had little difficulty in persuading the '66' Company's directors that the new construction should represent a significant advancement over her predecessors — particularly in the interior design. By this stage, Poul Kjærgaard had left Kay Fisker to to start his own architectural firm (which still exists today) and his ship-designing role was taken over by other assistants — Aage Nielsen, Robert Dueland Mortensen and Mogens Didriksen. Kjærgaard, however, remained in close contact with Fisker and his former colleagues and, perhaps understandably, took a keen interest in the *Kongedybet*'s design and outfitting. He later recalled that:

'The [*Kongedybet*] featured a couple of notable innovations. Externally, the funnel and the front of the superstructure were made more curvaceous and harmonious [than on previous '66' Company ships]. With this outward appearance at the front, it was possible to plan the smoking saloon within a regular ellipse shape. Fisker visually strengthened this clear geometric form through a correspondingly clear

The *Kongedybet*'s first class saloon was an inspired design with an abstract ceiling featuring concealed lighting and also the ventilation system. (courtesy of Pauli Wulff of Poul Kjærgaard Architects)

grouping of the furniture, while the ceiling design made the room even more interesting. A composition of large abstract arabesque cut-outs both increased the ceiling height and gave more room for indirect lighting and ventilation. With such ships now fitted with stabilisers, it was possible to design furniture for the various saloons much lighter than in the earlier vessels and *Kongedybet*'s smoking room benefited in particular from this innovation.'[20]

The *Kongedybet* was delivered from B&W and introduced in April 1952. At 2,314 grt, the new '66' Company flagship was the largest yet seen on the route, yet her popularity led to her being returned to her builder in 1958 for lengthening. The hull

The Kongedybet catches evening sunlight as she sails through The Sound en-route to Bornholm. (courtesy of Pauli Wulff of Poul Kjærgaard Architects)

and superstructure were cut in half and a new mid-section was inserted.

The *Kongedybet* continued in service to Bornholm until the late-1970s. In 1973, the '66' Company was taken over by the Danish Government, which created a new firm, called Bornholmstrafikken, to run services to the island. In 1979, the *Kongedybet* was replaced by the first of two new car ferries. Sold to the Government of the People's Republic of China, she was re-named *Baopeng* and sailed thereafter in the Chinese coastal trade between Hong Kong and Shantou. Later renamed *Ding Hu*, the vessel was broken up in Hong Kong in 1986.

The *Kongedybet* was to be Fisker's last ship design project. By then, he was approaching sixty and was involved in many building projects on *terra firma*. Moreover, the younger generation of Danish architects, many of whom he had taught at Charlottenborg in the 1930s and 40s, had become interested in the theories of the influential German modernist architect and theoretician, Ludwig Mies van der Rohe. Mies had by then become Professor of Architecture at the Illinois Institute of Technology in Chicago. His strictly formal and constructionally determined architecture utilised completely open planning and emphasised the interplay of interior and exterior. The photographer, Keld Helmer Petersen recalls that 'when Danish architects of the younger post-war generation went to Chicago in the early-1950s and saw Mies's Lake Shore Drive Apartments, his IIT campus and the Farnsworth House, many thought that there was little more left to be said.'[21] Young architects, such as Kay Kørbing, moved modern Danish architecture forward, essentially by blending the sophistication and technical refinement of Mies with Fisker's belief in 'the functional tradition', producing designs at once more decisively modern, yet still making the best use of traditional Danish building materials.

Glossy wood veneers predominated in the *Kongedybet*'s first class hallway. Note that the exit doors are studded with small circular windows – perhaps a rare concession by Fisker to 1950s design taste. (courtesy of Pauli Wulff of Poul Kjærgaard Architects)

THE WORK OF KAY FISKER AND KAY KØRBING

SHIPS DESIGNED BY PALLE SUENSON

Between Kay Fisker's final ship designs and Kay Kørbing's first, a small number of ships were designed for DFDS and other companies by another distinguished Danish architect – Palle Suenson. Suenson was born in 1904 and so was of a younger generation than Fisker. He shared a very similar design philosophy, however, producing elegantly proportioned housing and commercial premises in a progressive interpretation of Danish traditions. Suenson's first ship interior design project was the post-war refitting of the Bergenske Dampskibs Selskab's *Venus* in 1946. This ship, which had been designed by the young Knud E. Hansen in 1931 (see above), was seized by the Germans during the occupation of Norway and was taken to Hamburg where the Allies found her in a sunken state, with the bow shattered by a bomb explosion, in the Spring of 1945. She was salvaged, however, and sent back to the builders' yard at Helsingør for a three-year restoration and modernisation from which she emerged with a much changed exterior profile and new interiors, very much in the manner of Kay Fisker's pre-war work.

Later, in the early-1950s, Suenson was invited by DFDS to design interiors for its two new passenger ships then under construction for the popular domestic overnight service from Copenhagen to Aalborg in Northern Jutland. As with previous ships for the company, the *Jens Bang* (named after a 16th century Aalborg merchant) and *H.P. Prior* (the owner of one of the original constituent companies which had formed DFDS) were ordered from the Helsingør Skibsværft. Suenson's approach to shipboard design was clearly inspired by Fisker. He too favoured dark wood panelling and a clear uniformity of design throughout the passenger accommodations. However, while Fisker's interior designs tended to be smooth and streamlined, by this point Suenson advocated a more constructionally expressive approach, emphasising the ships' structural framing, which was encased in wood veneer to make a repeated motif of strong vertical mullions from floor to ceiling between the windows. This use of repeated vertical elements could also be found in his later architectural endeavours – such as his highly regarded administration building for the F.L. Smidth Company in Copenhagen (1956). As we shall see, several of Kay Kørbing's ship interiors expressed the framing in a similar manner.

The *Jens Bang* was delivered to DFDS in March 1950 and the *H.P. Prior* was delivered three months later. These two attractive ships, each of 3284 gross tonnes,

The *H.P. Prior* at sea, en-route through the Kattegatt from Copenhagen to Aalborg. (Author's Collection)

The *Jens Bang* in the 1960s after slight rebuilding of the aft superstructure to increase her capacity. (Author's collection)

could carry 86 first class passengers, all in comfortable cabins, 224 second class passengers and 1190 unberthed in deck class. As the route was very popular, they had to withstand punishing use with many passengers staying up all night, or sleeping wherever they could find a corner in the saloons and hallways. Certainly, the atmosphere onboard was very different from that on DFDS's longer international routes.

Palle Suenson's details for these ships were re-used many times by Helsingør Skibsværft for the interiors of subsequent passenger vessels built there, such as the French train ferry *Saint-Germain* (1951) and the Portuguese passenger liner *Funchal* (1963). Although Suenson designed few ship interiors and this work was peripheral to his practice, his designs are significant, not only because of their outstanding quality of materials and detailing, but also because they represent an intriguing evolutionary link between the points at which Kay Fisker left off and Kay Kørbing continued.

The *Jens Bang*'s smoking saloon shows Palle Suenson's approach to shipboard design with rather more angular furniture, vertical structural members encased between the windows and indirect lighting reflected off the smooth white ceiling. (Author's collection)

CHAPTER 2
SHIPS DESIGNED BY KAY KØRBING

At the The Royal Academy's Architecture School, one of Professor Kay Fisker's students was Kay Kørbing who went on to become the best known Danish architect involved in ship design.

Kørbing was born in Copenhagen in 1915, the son of J.A. Kørbing, who was the technical director of DFDS from 1921, its managing director from 1935 until 1955, and, subsequently, its chairman. Kay Kørbing initially planned to follow in his father's footsteps but, perhaps fearing claims of nepotism or possible conflicts of interest, Kørbing senior did not encourage his son to become involved with shipping. He worked instead as a bricklayer in Copenhagen – a beginning which he says first got him so interested in the raw materials of building and which taught him the importance of good architectural detailing. His interest in design led him to join first Det Tekniske Selskabs Skole in 1938 to gain admittance to the Royal Academy's Architecture School, where he studied until 1942. Under Professor Fisker, there was a hothouse atmosphere with many talented students – Kørbing was in the same class as Jørn Utzon, later the architect of the Sydney Opera House. He recalls Fisker as 'very haughty and self-assured, but also a very inspiring teacher. He was certainly a busy man, darting between the architecture school, his own office, various building sites and the Danish Architectural Institute's headquarters.' When Kørbing graduated in 1942, Denmark was under German occupation and work was scarce, and so he left for Sweden, finding employment with the highly respected architect and urban planner Cyrillus Johanson. There, he produced the competition-winning design for a church in the Storä Essingen district of Stockholm.

The Storä Essingen Kyrka was actually completed after the war, by which time Kørbing had returned to Denmark and so another of Johanson's assistants, by name of Asplund, supervised its construction.[22] Located at the crest of a hill, it is a delightful modern romantic composition in beautifully crafted red brickwork, dominated by a free-standing bell tower. The interior is calm and meditative with judicious attention to the lighting and other small-scale details. Evidently, Kørbing's earliest work demonstrates a compositional flair and a feeling for Scandinavian building and craft traditions which remained evident throughout his later career in ship interior design.

Kørbing returned to liberated Denmark in 1945 and his first job was the rebuilding and interior design of Denmark House in London's Piccadilly, the headquarters of the Danish Tourist Office. Thereafter, his first ship design was a smoking room for the DFDS cargo liner *Naxos*, built in Frederikshavn in 1955, which could carry 12 passengers. Kørbing remembers how he got the job:
'It so happened that my godfather was a director of the Helsingør Skibsværft which designed everything for DFDS. He tried several architects and then asked me to do some sketches because 'you know your father's taste.' I initially declined, but one of his fellow directors, a man called Garde, persuaded me and I did some sketches. As it happened, my father liked the drawings and he asked Garde 'who drew these?' 'Your son!'"[23]
Next Kørbing designed the cabins and saloons for two DFDS passenger and cargo liners for the North Atlantic trade – the *Oklahoma* and the *Ohio*, both built at Helsingør and delivered in 1956.

MS *PRINSESSE MARGRETHE* AND MS *KONG OLAV V*

After the cargo liners, Børbing was asked to design the passenger vessel *Prinsesse Margrethe*, another fine ship for the Copenhagen-Oslo route. 'My godfather asked me to do general arrangement drawings for the new Oslo ship 'and we'll do the rest', but knowing how important good detail was, I refused unless I could do all the interiors and design them down to the smallest elements. I soon found that shipyards don't really like architects as, to them, we are an unpredictable third party coming between them and the shipowners, so I quickly learned that it was necessary to be firm, but cajoling to persuade everybody to go with my proposals... 'Of course, I was heavily influenced by Kay Fisker's teaching, so the starting point for the design had to be the existing DFDS ships he had designed. Having worked on the interior of Denmark House in London, I was familiar with the *Kronprins Frederik* and *Kronprinsesse Ingrid*, which were very beautifully designed, so I used them as references for my own work. However, I was also very determined to be my own man, so with what might be called youthful enthusiasm, I analysed all the aspects of the design and outfitting of such ships to find out what the design possibilities were. That both my father and godfather were in senior positions in DFDS certainly helped as I got business advice from one and detailed technical information from the other. This enabled me to be really quite clinical in my analysis of the design problems, so I think that the *Prinsesse Margrethe* represented a big leap forward in design terms.'[24]

The engine room was located two-thirds aft and the exhaust uptake was infact routed through the ship's after mast. The funnel, slightly forward of amidships, actually contained the air conditioning plant. Thus, the first class public rooms, which occupied the forward half of 'A' deck were well away from any engine noise. Aft, the second class cafeteria (also on 'A' deck) and the dining saloon, on 'B' deck below, were divided into two smaller sections by the casing surrounding the exhaust uptake. This was a most ingenious layout and, indeed, the *Prinsesse Margrethe* was

The brand new and very sleek *Prinsesse Margrethe*, shortly after entering service, is seen at speed in the Øresund in this DFDS publicity photograph (Author's collection)

The handsome *Kong Olav V* (Paul Clegg)

perhaps the ultimate development of the small motor liner for overnight service before the advent of car ferries. According to Kay Kørbing:

'To give the most spacious feeling to what was by today's standards a fairly small ship, I decided to create 'service towers' through the different decks to contain the plumbing, the wiring and stair cases. These were located well within the shell of the ship, so you could see all around them. The *Prinsesse Margrethe* had truly open plan interiors and we had to work carefully round the existing design regulations. I wanted fully glazed bulkheads between the saloons to make a greater feeling of space, so we developed retractable fire doors which folded away into the ends of the service towers. When you looked along the length of the ship, there was a

Forward on the *Prinsesse Margrethe*'s saloon deck was the spacious U-shaped first class saloon. (Keld Helmer Petersen)

On either beam were the first class cocktail bar to starboard and the rear section of the smoking saloon to port, shown here. This featured a black and white wall design by Ole Schwalbe. (Keld Helmer Petersen)

continuous vista from one room to the next and on the end walls, there were specially commissioned abstract murals by Danish artists whom I knew. I also designed a whole new range of furniture for the ship. The modern furniture you could buy in Denmark at the time was very refined, but hopeless for ships as it would tip over in bad weather and the ship furniture that was available was usually so heavy and old fashioned. The moulded fibreglass cafeteria chair I designed for the ship was just about the first of its kind anywhere. Everything had to be good-looking but very hard wearing and easily maintained. I also did tubular steel-framed chairs for the

The first class entrance hall featured a reception desk and a Y-shaped staircase. Even in first class, the ship was rational, yet stylish with easily maintained blue rubber flooring and brushed aluminium details. (Keld Helmer Petersen)

The first class dining saloon was U-shaped and was inserted around the galley and exhaust uptakes. This ensured that the majority of diners enjoyed a sea view. Similar in design to its second class counterpart, it featured and upholstered version of Kay Kørbing's glassfibre chair. Interestingly, there are no table cloths in this photograph by Keld Helmer Petersen as he thought that removing them would emphasise the functional quality of the design. (Keld Helmer Petersen)

lounges and special lighting was manufactured by Örrefors to give the right subdued effect.'[25]

One of the outstanding features of the *Prinsesse Margrethe,* and of subsequent ships designed by Kørbing, was the range of specially commissioned artworks integrated into her interior design.
'As a young man I was very interested in modern art and in the idea that architecture is 'the art of building.' Also, passenger ships have traditionally been floating showcases of culture and modernity, so I think it was natural that I invited artist-friends to produce artworks to adorn the bulkheads of the *Prinsesse Margrethe*. It was important that these pieces were not an afterthought, but were fully integrated into the overall design – for example, to finish off important vistas through the open-plan spaces, or to draw attention to particular facilities. I didn't want representational painting, but abstract pieces which not only reflected the modernity of the ship itself, but also might provoke curiosity among the passengers, giving them something to interpret for themselves...'[26]

Thus, appropriately enough, to the rear of the grand staircase in the first class hallway, the painter Harald Hansen made a large portrait of the young Prinsesse Margrethe. Young artists produced mural panels and bas-reliefs in the other principal public rooms. The end wall of the first class cocktail bar on the starboard side had a remarkable abstract mural by Gunnar Aagaard-Andersen while, facing the sea behind the bar counter itself was a bas-relief panel by the sculptor Svend Dalsgaard. On the port side bulkhead was a black and white painted panel by Ole Schwalbe. The first class dining saloon, aft of the hallway, had an iron bas-relief by Helge Holmshov, while the second class foyer had a striking black and white photo mural by the photographer Keld Helmer Petersen.
Colours were carefully selected and the judicious use of subtle shades of green, blue

DANISH SHIP DESIGN 1936-1991

The *Prinsesse Margrethe*'s second class cafeteria was the first self-service catering facility on a DFDS ship. Note how the ship's structural framing is expressed in the ceiling and wall finishes and how every seat outboard has a window and every table a lamp. (Keld Helmer Petersen)

and grey for furnishing in the forward saloon was beautifully accentuated by reflections of the sea and sky through the generous expanses of windows. Equal attention was paid to the atmosphere of the ship at night. Indeed, each armchair was placed below an individual ceiling light, while each grouping around the perimeter had its own table light. At night, these gave a soft glow which reflected warmly in the veneered wall panelling. While the flooring, wall finishes, lighting and furniture design were the same throughout, different palettes of colours were used to give each space a distinct character. The second class cafeteria, for example, had the same fittings as the first class dining saloon, except that its chairs were of bright red glassfibre, rather than red upholstery. The ship's design was explained in a special colour brochure, produced by DFDS, to commemorate the inauguration:

The same bulkhead, seen from the aft second class sundeck. (Keld Helmer Petersen)

'Form and function work together with light, hygeinic and aesthetically attractive materials, steel, plastic and rubber. The big entrance foyer, where the *Prinsesse Margrethe*'s stewardesses welcome first class passengers, is bright, fresh and airy. Thick glass partitions, Oregon pine and palisander, discreet lighting and soothing colours work together to give the feeling of spaciousness for which the ship is well-known.

The second class hallway had a glazed aft bulkhead at saloon deck level, giving a panoramic view over the ship's stern. (Keld Helmer Petersen)

Everything is arranged for travelling comfort and for the welfare of the guests. All possible desires have been thought of, both inside and out, so that the immediacy of the sea and coast complements the outstanding elegance within.'[27]

Introduced in 1957, the *Prinsesse Margrethe* was immediately hailed as being among the most beautiful of her type and was a great success, so much so that a sister ship was ordered from the Aalborg Vaerft for delivery in 1961. The *Kong Olav V* was equally popular, but DFDS had underestimated the growth in car traffic taking place in the 1960s and neither ship was designed to carry more than a couple of dozen cars, so within seven years they were replaced by larger car ferries with the same names. By this time, Kørbing had set up an office in Kvæsthusgade, close to the DFDS headquarters, but his expanding practice[28] was busy not only with ships, but also with the prestigious conversion of a series of courtyards off Strøget in medieval Copenhagen into Illums Bolighus – a designer department store specialising in modern Danish furniture and consumer goods.

Replaced by car ferries of the same names, the *Prinsesse Margrethe* and *Kong Olav V* were moved briefly to DFDS's secondary routes. After a couple of years sailing between Newcastle and Esbjerg with her name abbreviated to *Prinsessen*, the *Prinsesse Margrethe* was sold in 1971 to the Åland Islands-based Birka Line, becoming the *Prinsessan*. The *Kong Olav V* had her name shortened to *Olav* and was sold to Hong Kong owners in 1969, becoming the *Taiwan* and sailing between Hong Kong and Keelung. In 1972, the ship was back in Scandinavia and joined her sister in the Birka Line fleet as the *Baronessan*. Both operated 24-hour cruises to Mariehamn from Stockholm. In 1978 the *Prinsessan* was sold to Saudi Arabian owners and entered the Red Sea pilgrim trade as the *Wid*. The vessel was broken up in Pakistan in 1987. The *Baronessan*, meanwhile headed out East again after being sold to the Chinese Government. Until recently, she was in service from Hong Kong to the Chinese mainland as the *Nan Hu*, but was reportedly in very scruffy condition. It is thought that that the *Nan Hu* was scrapped by Chinese breakers in the late-1990s.

DANISH SHIP DESIGN 1936-1991

MS ENGLAND

The brand new *England* shows off her graceful lines in this photograph of the ship on trials immediately prior to her delivery to DFDS. (Author's collection)

As car traffic grew rapidly on DFDS's North Sea routes, the time-consuming crane-loading of vehicles into the holds of the *Kronprins Frederik* and *Kronprinsesse Ingrid* was no longer satisfactory, especially as the Danish government was avidly promoting Danish tourism in Britain. In 1962, DFDS went back to Helsingør Skibsværft to order a new passenger liner with drive-through car capacity. This was the 8,221-ton *England* which was to be innovative in many respects.

Another view of the *England*, this time showing her port side, whilst on trials in The Sound off Helsingør. (Author's collection)

A fine aerial view of the *England* during the mid-1960s; the cranes, mounted on her bow, were rarely used to load cargo, but instead lifted the car ramps into place to allow unloading through side doors in the shell plating. (Author's collection)

In 1960, DFDS had appointed an Englishman as its new Chief Superintendent Engineer – Bryan Corner-Walker. He was ex-Royal Navy and, as the Second World War ended for him with Denmark's liberation, he decided to stay in Copenhagen and to study naval architecture there. At DFDS, he was responsible for the design of the new passenger vessels of the 1960s and 70s, initially working under J.A. Kørbing. He recalls:

'Chairman Kørbing realised that growth in car traffic was becoming a significant trend and that new ships would be required. He was adamant that DFDS was not to have ferries, though, as he seemed to think such ships would be slightly demeaning to the company's traditions and, moreover, not sufficiently robust to cope with North Sea storms. Instead, he wanted a passenger liner which happened to carry cars as well. Comfort was a priority and I was asked to make the *England* as smooth-sailing

The *England*'s first class saloon developed the idea of a full-width space reflecting the form of the forward superstructure, as seen on previous DFDS ships designed by Fisker and Kørbing. (Author's collection)

as technology would allow. The ship evolved as a large, graceful motor yacht with very fine hull lines and she did indeed prove to be a most outstanding performer, even in the worst North Sea weather.'[29]

The *England*'s sleek hull design was developed from that of the Portugese-owned liner *Funchal*, completed at Helsingør the previous year. Two 14,000 horsepower B&W diesels gave a speed of 21 knots with reserve power to catch up in case of delays – very important in the treacherous North Sea. For Walker, safety was a

Uninterrupted expanses of glazing separated the saloon from the hallway and restaurant amidships, as shown in these DFDS publicity photographs. While fashions in clothing changed, the *England*'s modern interiors remained largely unaltered throughout her 18-year DFDS career. (Author's collection)

priority and, following the best British design practice, the *England* was fitted with an automatic sprinkler system to deal quickly with any outbreak of fire. Such thorough fire protection was almost unheard of on most European and American ships at that time. The Italians and Americans, in particular, preferred to use fire-resistant interior finishes, rather than sprinklers, believing that prevention was better than cure, but such a policy could not protect against kitchen fires or burning soft furnishings. Moreover, the desire for increased open planning of interiors made active fire protection even more desirable and the destruction by fire of the *Kronprins Frederik* just over a decade before must have been at the forefront of the DFDS directors' minds (see above).

Early on in the design process, Kay Kørbing was appointed to design the new vessel's interiors:
'I was very pleased to get the job of designing the *England* as this really was to be something special in the Danish fleet – a floating first class luxury hotel. Ships of this kind are designed around their passengers, so the starting point for the whole process was the interior. Ship interiors tend to be designed in multiples of 210cm, the width of a table with a chair on either side and space to pull each chair back. Conveniently, this dimension is also the width of a typical cabin. Every table and every outside cabin needs to have a window, so this governs where the vertical structural members go. All of this was carefully thought out on the *England* to make the ship as enjoyable from the passengers' point of view as possible.
In appearance, the *England* was very elegant and modern. I remember a meeting in DFDS's boardroom where company directors and technical people, all the design team from the shipyard and my interior architects sat around a mock-up model of the ship. Bryan Corner-Walker pulled this carefully shaped lump of wood from his bag and stuck it on top of the model. 'How would that be as a funnel?' It was streamlined and with fins to stop the smoke being sucked down onto the deck. Everyone agreed that it looked good and it was a design we used subsequently on other DFDS

The second class public rooms were of a similar quality to their first class counterparts, as shown by the smoking saloon, located aft. Following DFDS's decision to make all of its ships one class in the early-1970s, this was re-named the Hansa Bar. (Author's collection)

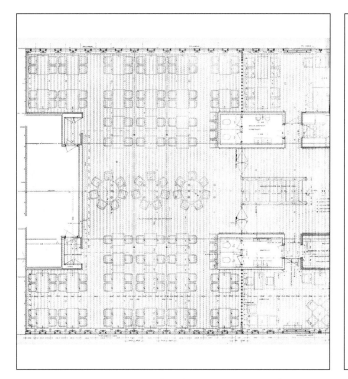
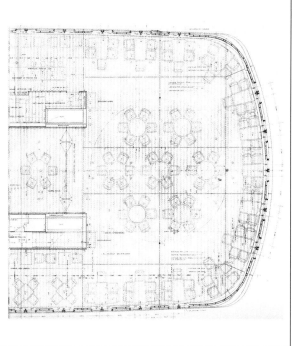

Detailed general arrangement drawings by Kay Kørbing for the England's restaurant (left) and first class saloon (right). The seating arrangement is clearly related to the window locations and the lighting plan allows every seat to be lit individually. (Kay Kørbing)

passenger ships – in fact it was also used as a company trademark in publicity material.'[30]

Not only was the *England* a great beauty, she was also one of the best-appointed ships of her type and it was rumoured that she would be used on cruises as well as for ferry work. Some 155 first class passengers and 244 second class passengers plus 100 cars, loaded through doors in the sides of the hull, could be carried. Forward

The first class sun deck above the bridge in its short-lived original form before it was enclosed with glassfibre panels. (Author's collection)

on the saloon deck was the first class smoking room, the interior form of which followed the curved and slanted shape of the superstructure. It was panelled in teak and, as with all of the passenger spaces, it was fully air-conditioned with ribbed ceilings specially designed to blow in cool air and extract stale air all over to keep the ship fresh and evenly heated. The furniture was again specially designed with a new range of distinctive cubist lounge chairs, manufactured by Thorballs Eftf, and light fittings with Örrefors glass, made by Lyfa, set into the ceiling to cast a warm glow. The furniture in this large space was ingeniously laid out so as to create an inward focus by night with the seating and lighting in the middle forming a circular composition. Around the perimeter, groups of chairs were placed around rectangular tables at each window bay. Kørbing repeated this subtle, but effective, arrangement on other ships in the ensuing years.

To the rear of the first class section was the first class dining saloon, with a decorative panel in yellow by Arne L. Hansen. In between was a hallway with floor-to-ceiling plate glass bulkheads and doors without frames to give an uninterrupted expanse of glazing. On the port side, there was a copper panelled cocktail bar adjoining the smoking room and to starboard, a writing room.

The first and second class areas were separated by the galley, which served both restaurants – incidentally the one in second class was very similar to its first class counterpart. Its central feature was a relief panel depicting the sun's rays over the sea by Rolf Middelboe. Overlooking the stern and the second class sun deck was the second class smoking room, which was light and airy thanks to floor-to-ceiling windows in the rear bulkhead. One deck below were the *England*'s first class cabins, shops and a playroom – a first on a North Sea passenger ship which showed DFDS's belief that, no matter what age, everyone should enjoy their trip. The *England* had covered promenades on either beam and extensive teak sun decks. The first class sun deck was above the bridge and was sheltered by glazed screens to allow passengers to see the ship docking – always a fascinating experience. In stormy weather, however, the spray sometimes came right over the *England*'s superstructure, and so the space was soon fully enclosed with a glass fibre roof, becoming a kind of winter garden.

Unlike the Fisker-designed ships in which DFDS had insisted on using its own, rather perfunctory, cabin designs for second class, in the *England*, Kay Kørbing devised his own range of modular cabin furniture, which could be used in various permutations to suit different shapes of room. His solution was ingenious, but simple. Dados, veneered in African wengé (a dark hardwood) were pre-fitted with electric sockets, light switches and the mountings for the berths. These were cut and fitted to the lower walls of the cabin spaces, after which the remaining fixtures could simply be hung in place. The upper walls and ceiling were painted matt white.

Spacious, undeniably luxurious and fully stabilised, the *England* brought a new level of quality to short-sea passenger ship design and effectively made Kay Kørbing's reputation in Scandinavia and beyond as a highly talented architect of ship interiors. On 4th June 1964, the *England* left Harwich on her maiden trip to Esbjerg and, after a short period in service, the delighted DFDS directors quickly turned their attention to building a sister ship (see below). By this time, the old shipping company had had a boardroom shake-up after the J. Lauritzen Rederi bought a substantial shareholding in 1964 and Knud Lauritzen of that firm replaced the elderly J.A. Kørbing as chairman.

The *England* remained on the Harwich-Esbjerg route until 1974, when the larger

The *England*'s cabins were semi-prefabricated and were a design innovation of which Kay Kørbing was particularly proud. The lower wall panels and furniture could be inserted into the cabin spaces to give standard fully fitted out interiors relatively quickly and inexpensively. This is a first class cabin on the promenade (main) deck. Subsequent DFDS ships of the 1960s also utilised this design. (Keld Helmer Petersen)

Dana Regina displaced her to the seasonal Newcastle-Esbjerg route (see below). Latterly, her Achilles heel had been her lack of freight capacity as her garage was only one deck high. She was laid up at the end of the 1982 summer season and sold the following year to Cunard for service between Cape Town and Port Stanley to carry workers and equipment involved in building the Falkland Islands airport. After a spell in lay-up at Birkenhead, she was sold in 1986 to the Greek oil and property tycoon John S. Latsis who at first used the ship in the Red Sea as the *America XIII* and, later, as the *Emma*. Latterly laid up at Eleusis and totally gutted for conversion to a private yacht, the project was eventually abandoned. Named *Europe*, but reduced to an empty rusted shell, the vessel was eventually sold for scrap, but sank in the Red Sea en-route to the ship breakers in April 2001.

The gutted remains of the *Europe* (ex *England*) at anchor in Eleusis Bay in Greece after John Latsis aborted his conversion of the ship to a super yacht. (Peter Knego)

TERMINAL BUILDINGS

Kay Kørbing's architecture on *terra firma* is characterised by its sobre, rational appearance, its use of high-quality materials and its fine detailing. Shortly after the introduction of the *England*, he was commissioned by DFDS to design a number of new passenger terminal buildings. The first, in Federikshavn (1965) was a single-storey structure which utilised a glue-laminated timber roof construction system upon cast concrete walls. This was followed by a combined terminal, railway station and warehouse complex in Esbjerg (1966-7). It was a spacious three-storey edifice with a reinforced concrete frame and brown brick infill. Facing the harbour on the upper floor were tall windows, separated by slender mullions with a viewing gallery in front – the perfect spot from which to contemplate the ships tied up along the quaysides. Finally, Kørbing designed a further DFDS terminal building in Aalborg (1968), again in concrete and brick. All remain extant and in near-original condition, although the Aalborg building has been converted to offices and the one in Esbjerg has recently been vacated by DFDS and is presently threatened with demolition to make way for new housing.

FURNITURE

The range chairs designed by Kay Kørbing for the *England*, consisting of KK7 restaurant and KK8 lounge chairs and setees, was of such high quality that they were later sold commercially.[31] Initially exclusively marketed in Denmark by Godtfred H. Petersen of Store Heddinge, from 1966 onwards they were also manufactured and sold under license in the United States. At that point, Kørbing entered into an unusual and innovative agreement with Morris Goldman of the New York-based JG Furniture Company to fabricate slightly revised versions of his

The new DFDS terminal building in Esbjerg, viewed from the boat deck of the *Winston Churchill* on her maiden arrival in 1967 (Kay Kørbing)

DANISH SHIP DESIGN 1936-1991

On its landward side, a long canopy connected the DFDS terminal with the platforms of Esbjerg Havn railway station and enabled passengers to make an easy connection from the 'Englanderen' boat train to and from Copenhagen (Author)

furniture in return for a royalty payment on each sale. This deal also involved the Danish fabric and furniture designer Nanna Ditzel, who produced a range of rugs and fabrics, also to be manufactured by JG Furniture, to complement Kørbing's furniture. Unfortunately for Kørbing, another unnamed American furniture manufacturer bought examples of the chairs made by JG and produced their own version without either asking permission or paying any royalties. Subsequent chair designs by Kørbing were manufactured in Denmark only and were retailed by Godtfred H. Petersen until that firm closed down in the mid-1990s.

Kay Kørbing's glassfibre cafeteria chair, used in most of his ship designs from 1957 until 1991 has become something of a Danish design classic. His lounge and restaurant furniture was equally practical and hard wearing and shares a family resemblance with chairs of the same period by other Danish designers, such as Arne Jacobsen, Poul Kjærholm and Hans J. Wegner.

MS *SAGAFJORD*

As with all of the post-war Norwegian America liners designed by the company's own technical department under Kaare Haug, the *Sagafjord* was an outstandingly beautiful modern liner. (Bård Kolltveit)

Building on his growing reputation and on the renown of the *England,* Kørbing's next shipboard project was a most prestigious commission. Responding to the jet era, which was undermining Norwegian America Line's (NAL) transatlantic passenger trade, in 1960 its directorate instructed their technical department to draw up specifications for a luxurious passenger liner; this was to serve both transAtlantic and cruising purposes. Due to an international conference treaty between the transAtlantic lines, the ship would have to be capable of being subdivided into two classes, carrying 800 passengers for the Oslo-Copenhagen-Stavanger-New York run with a capacity of 450 in a single class when cruising.

The majority of the cabins were to be outside staterooms and all were to have showers and toilets. In her role as a cruise ship, all passengers were to be accommodated in beds, not bunks, and both the ballrooms and dining rooms were required to have sufficient capacity to accommodate all cruise passengers at a single leisurely sitting. This was a tall order. Air conditioning was to be installed throughout for both passengers and crew, and both were to be provided with outside swimming pools, as well as an indoor pool. As with more recent Norwegian America Line predecessors, the *Oslofjord* and the *Bergensfjord,* the new ship was to be a

Viewed side-on and at speed, the *Sagafjord*'s flowing lines are seen at their best in this late-1960s publicity photograph for Norwegian America Line. The shapely silhouette was later spoiled by the addition of a further deck of penthouse cabins above the bridge. (Michael Zell collection)

twin-screw motor vessel with an all-welded aluminium superstructure. Norwegian America Line's own chief naval architect Kaare Haug once again was in charge of the new ship's technical design.

The specifications for the new liner, requiring spacious public rooms and cabins, meant that the dimensions would have to be considerably increased in comparison with previous NAL passenger vessels, although the width was restricted to only 80 feet because of the drydocking facilities in Oslo. The same basic hull form as the earlier ships was chosen and in May 1962 model tests were carried out at the Norwegian Technical High School in Trondheim. In September 1962, a contract for the construction of the ship was signed with the French shipbuilders, Société des Forges et Chantiers de la Méditerranée, La Seyne, for delivery in 1965. The keel was laid in May 1963 and later that year it was announced that the name for the new ship was to be the *Sagafjord*.

The interior design of the *Sagafjord* was to represent a radical change in direction for NAL. In line with the upsurge in Norwegian national sentiment following liberation from the Germans, the company's initial post-war ships had been in the national romantic idiom with interiors largely designed by Arnstein Arneberg, the architect of Oslo City Hall. To make the *Sagafjord* more appealing to the wealthy, cosmopolitan

The forward stairwell featured an impressive gently curving stairway in an oval void cutting through the full height of the ship (left). First class cabins were exceptionally spacious and the suites designed by Kay Kørbing were especially fine with separate seating areas and full length curtains (right). (Keld Helmer Petersen)

The North Cape Bar was located on the *Sagafjord*'s port side. The space was panelled in teak with copper mosaic tiles behind the bar and black leather upholstered chairs. With its large windows giving fine views of the ship's progress, it was a favourite venue for pre-dinner drinks. (Keld Helmer Petersen)

clientele Norwegian America's directors hoped to attract, they assembled an international team of architects and designers under the co-ordination of the Norwegian firm, F.S. Platou. The French architect Georges Peynet designed the theatre (Peynet had previously designed the theatre of the French Line's giant SS *France*). The most spacious room on the ship was the tourist class ballroom, designed by the Norwegian Finn Nilsson. With its high ceiling and a floor area covering 8,000 square feet, a bright red colour scheme and subdued lighting, it was commodious, yet instantly welcoming. From this room, glass doors led out onto the lido deck and open-air swimming pool which had an illuminated fountain at night. One deck above was the Polaris Night Club, the work of Han VanTienhoven who was also responsible for the main entrance hall. The magnificent Saga Dining Room seated 468 and was designed by F.S. Platou and his assistant, Njål Eide. Its central section was two decks high and, descending a splendid staircase leading down from the deck above, passengers made a grand entrance through doors in a double height fully-glazed bulkhead. Everywhere there was the air of quiet luxury and spaciousness with large cabins (90 per cent of which were outside rooms) and wide, uncluttered corridors.

Kay Kørbing was put in charge of the design of the first class accommodation, at the forward end of the veranda deck. At his suggestion, and following the precedent of the *England*, the traditional transAtlantic liner concept of a wraparound enclosed promenade was abandoned in favour of allowing the public rooms to occupy the full width of the ship with large windows overlooking the sea.

Externally and within, the *Sagafjord* was a ship of sleekly animated curves, great expanses of windows, grand sweeping staircases and tapering lines. Consequently, Kay Kørbing's interior designs reflected upon and accentuated this vocabulary. For example, the circular Garden Lounge, located forward, was built on split levels around a sunken circular dance floor and ingeniously combined the daytime role of a quiet sitting room with that of an intimate nightclub. This was achieved by means of four vertically slatted matt gold metal screens, which gave the space a cosy and warm-toned inner area and a bright and airy perimeter. The ceiling above the central section featured lighting coves with a colour-change system for use at night when there was lounge dancing. Elsewhere, many of the same fittings were used as in the *England*. The first class hallway, immediately behind, was an impressive oval-shaped atrium with a dramatic under-lit open tread staircase plunging down the full height of the ship and appearing to float in front of a gently curving wall, panelled in teak and onyx marble. Given the increasing sheer and camber of the ship towards the bow, the calculation of the dimensions and the immaculate detailing of this structure were considerable achievements in themselves.

Moving further aft, the library, writing room and North Cape Bar were on either beam. The library took the form of a compact semi-circle of book shelving with comfortable armchairs and a large illuminated globe as its centrepiece. In contrast, the writing room, adjacent, was formally laid out with rows of desks on either side of a centre aisle. To port, the North Cape Bar was rich in tone with Canadian rock elm wall panelling, copper mosaic behind the bar counter, black leather upholstered armchairs and wood carvings by Dagfin Werenskiold in illuminated wall niches. By day, it was well lit through large windows, making it the ideal spot for pre-dinner cocktails or simply watching the sea going past, but full-length curtains ensured that it was dark and sociable by night.

Throughout the interior, contemporary Scandinavian, Dutch and French artists adorned the bulkheads and stairwells with specially commissioned installations, doubtless reflecting the tastes of the designers of these spaces. For example, an iron bas-relief depicting a fleet of Viking ships, finished in gold on a brilliant red background by Carl B. Gunnarson formed a backdrop to the grand staircase leading

The *Sagafjord*'s writing room was located on the starboard side. The inboard walls were decorated with laminated maps and there was a circular library space in the forward section (not visible in this photograph). (Keld Helmer Petersen)

Combining the roles of airy winter garden by day with intimate nightclub by night, the Garden Lounge was a space of which Kay Kørbing was especially proud. Its inner nightclub area combined rich colours, concealed lighting and a marble dance floor. (Keld Helmer Petersen)

into the Saga Dining Room, while Phyllis Evensen created a series of glass mosaics in the forward staircase on Sun Deck and the mural in the main Ball Room, entitled 'Water Music', was painted by Torstein Rittun. Other artworks aboard were by Sigurd Winge, Van den Broek, Mandaroux and Bongard.

In addition to the public rooms and circulation spaces, Kørbing also designed a number of first class luxury suites. He recalls that:

'The *Sagafjord* was among the most beautiful ships I was involved in. On an overnight vessel, every corner of space counts, but as this was a luxury liner, I could afford to be more generous. Compared with what the Italians and the French were building at that time, she was quite a modest ship, though. The design journals were full of articles about the *France* and the *Leonardo Da Vinci* and how wonderful they were, but I think that the *Sagafjord*'s owners got much better value for money overall and every detail was carefully thought out by the design team. One of my earliest memories of the project was a cinema show given to all the architects involved by Kaare Haug, showing an Atlantic storm filmed on one of the existing Norwegian America ships. Haug made it very clear that everything we designed would have to survive extreme weather conditions and considerable pitching and rolling.

'I designed a new version of my lounge chair for the Garden Lounge with a wicker seat and back to give the feeling of a modern interpretation of the traditional ocean liner winter garden. Thoughout, very opulent and solid materials were used – fine hardwoods, marble and black leather. Whenever metal finishes were required, I always insisted on matt treatments. On no account was the *Sagafjord* to be glitzy, instead it was to be rich, yet understated. I think we succeeded – certainly the French builders did a very precise job and I was very proud of the *Sagafjord*.'[32]

The *Sagafjord* was eventually delivered on 18th September 1965, some six months behind schedule, and arrived in Oslo on 24th September. At 24,000 tons, the liner was considerably larger than NAL's earlier ships, although she had the same unmistakable streamlined profile with a higher, longer superstructure and a single mast. Undoubtedly one of the most beautiful ocean liners ever built, she was renowned for her long, flowing lines and remarkable consistency of forms and details. Behind the long, flared bow, the superstructure was piled up in receding tiers with swept back bridge wings and an imposing tapering funnel, the profile of which matched the sweeping lines of the aft sun decks. The *Sagafjord* sailed on her maiden transatlantic voyage from Oslo on 2nd October 1965 and on 10th November left New York on her first cruise to the West Indies. This liner soon established itself as one of the finest and most popular cruise ships afloat and spent an increasing part of the time in this role.

In 1980, the *Sagafjord* was transferred to a newly-formed company, Norwegian America Cruises. This was 90 per cent owned by Norwegian America Line with the remaining 10 per cent of shares belonging to Leif Hoegh, another prominent Norwegian ship owner. In March 1981, Hoegh bought the Norwegian America Line outright, principally to gain control of its freight operations. Subsequently, Norwegian America Cruises was sold to Cunard Line in 1983.

Unfortunately, while cruising in the South China Sea in 1996, the *Sagafjord* suffered a serious engine room fire. The vessel might well have been repaired, but the financially ailing Trafalgar House Group, Cunard's parent company and a business more interested in the construction industry than shipping, was taken over by the Norwegian Kvaerner engineering group. Kvaerner wanted to sell Cunard and so the *Sagafjord* was set aside while negotiations took place. Fortunately, a charterer was found in the form of the German travel agency, Transocean Tours, who, having failed to secure the *Regent Sea* (ex *Gripsholm*) when her Greek owners went bankrupt and the ship was arrested, instead requested the *Sagafjord*. Prior to her arrest, the *Regent Sea* had already been marketed by the Germans under her original Swedish name. The *Sagafjord* was repaired and was herself then re-named *Gripsholm* to fill the gap.

After the charter ended, the vessel was put up for sale and was bought by a British company, Saga Holidays, which specialises in cruises for people aged over fifty. After a refit, the ship became the *Saga Rose* and now cruises mainly from Dover and Southampton. Over the years, her magnificent interiors have been altered better to suit conservative British tastes. Nevertheless the *Saga Rose* still remains one of the most pristine and elegant cruise liners in service today.

DANISH SHIP DESIGN 1936-1991

MS *WINSTON CHURCHILL*

The handsome *Winston Churchill* glides serenely through the limpid Mediterranean during her sea trials. (Author's collection)

In the mid-1960s, DFDS was approaching its centenary and celebrating in style with an unprecedented expansion and a new building programme. At that time, its own shipyard at Helsingør was fully occupied building a series of large cargo liners for the company's North Atlantic services and could not meet a quick delivery time, and so the next acceptable bid was from an unusual source – the Italian shipyard of Cantieri Navali del Tirreno e Riuniti at Riva Trigoso near Genoa. This family-owned yard built ships literally on the beach and launched them in almost complete condition and it was there that the keel for what would turn out to be one of DFDS's most popular and long-serving ships was laid in January 1966. Bryan Corner-Walker recalls that DFDS itself carried out the initial design work for these ships and drew the contract drawings (the naval architect dealing with this process was called Modeweg-Hansen). Once the contract was signed, the shipyard then refined these designs using its own technical experise. Dr Andrea Ginnante joined the yard as the DFDS ships were nearing completion:

'In the 1960s, the Italian shipyards could call on the skills of the most innovative and talented naval architects and engineers in the world. Here at Riva Trigoso, our naval architect was a brilliant man called Giò Melodia, who helped to refine the very beautiful and efficient hull forms for the DFDS ships. Undoubtedly, he was influenced by Nicolò Costanzi, who designed the legendary *Guglielmo Marconi*, *Galileo Galilei* and *Eugenio C.* Like the great Italian liners of the period, their lines at the bow were like a wineglass – concave at the waterline to cut through the waves efficiently and convex above to maximise the deck space. That elegant shape made these ships very distinctive.'[33]

The new ship's hull form was designed to be loaded with cars through the bow and stern, whereas the *England* loaded from the side. There were lifting bow and stern visors designed to continue the shape of the hull so that, when they were closed, no one would have known that the ship was a car ferry. The hull also contained an extra deck so that the car deck could be double-height to carry lorries and buses. Kay Kørbing recalls the building process:

'DFDS actually ordered a whole series of ships from the Italians – five passenger vessels and two bulk carriers. I was initially only given the job of designing the first of these new ships, an improved and enlarged sister to the *England*.

'I was sent to work in Genoa for three years as the new ship was being designed there. With Bryan Corner-Walker, I flew to Milan, then drove to Genoa every three weeks or so to keep an eye on progress. DFDS's technical staff wanted a ship looking outwardly similar to *England*, but the Italians, who have a great sense of style, believed that a more fundamental approach was required. My contribution was the interior design, which was indeed similar to *England*, but slightly more colourful, in line with late-1960s taste. I think the blend of Italian naval architecture and Danish interior design was a great combination and the new ship had immense style…'[34]

The new DFDS flagship was launched as the *Winston Churchill* on 25th May 1967, and the DFDS newsletter, DFDS Express, proclaimed 'A Great Ship Takes To The Sea'. The *Winston Churchill* was indeed a great liner in miniature, being one of the best

The *Winston Churchill*'s graceful bow towers above the slipway before the launching at Riva Trigoso in the Spring of 1967. This view shows the typically Italian design with a swan neck-shaped prow, similar to many of the most famous Italian transAtlantic liners. (Author's collection)

DANISH SHIP DESIGN 1936-1991

Above: A fine views of the brand new *Winston Churchill* on her delivery voyage to DFDS at the start of a brilliant career during which she was to sail on the most important North Sea routes and even on cruises to the Baltic and Norwegian Fjords. (DFDS)

Below: The nightclub, located aft on B deck, had a copper and steel dance floor and bright orange furniture. The space was relatively short-lived as, later on, it was converted to couchette berths to increase the ship's capacity. (Author's collection)

appointed of her type. After sea trials, she left Genoa for Esbjerg, then sailed for Harwich where, after a brief call and berthing trials, she sailed up the Thames and dropped anchor at Greenwich, opposite the Royal Naval College. There, amid great celebrations, the ship was named by Baroness Churchill. The gleaming new *Winston Churchill* then set sail for Harwich and the start of a brilliant career during which she was to sail at different times to most of the important ports in Northern Europe.

Within, the *Winston Churchill* was similar in layout to the *England*, but she was very much an improved version in nearly every respect. According to the DFDS brochure published to commemorate the ship's inauguration:

The *Winston Churchill*'s first class saloon was similar to the one on *England*, but more richly toned with Italian rosewood panelling and tomato-red upholstery, designed by Nanna Ditzel. This image was taken by Kay Kørbing during the ship's maiden voyage from Esbjerg to Harwich. (Kay Kørbing)

Looking aft, the vista through the fully glazed partitions from the first class saloon, through the hallway, and to the restaurant was finished off with an impressive tapestry panel by Urup Jensen. There could be little doubt that the *Winston Churchill*'s first class accommodation was the most elegant yet seen on a North Sea passenger ship. (Author's collection)

'Nothing has been spared to ensure that the *Winston Churchill* is one of the finest ships in the world. From the laying of the keel, DFDS technicians have been carefully supervising the building of the ship. Numerous tests have been carried out with models and also on board the sister ship *England*, whose construction and excellent seaworthiness have been praised by shipping experts in many countries.
Every effort has been made to find the best possible solutions to the many problems that occur in the design of a ship. For example, although the cabins on the *England* are extremely elegant and comfortable, full-size mock-ups were made before the *Winston Churchill*'s cabin designs were finalised. If improvements could be made – then they were made!'

Certainly, the *Winston Churchill* was richer, more colourful and even more luxuriously appointed than her elder sister with greater use made of carpeting, curtains and contrasting dark veneers. The lounges and restaurants, for example, were panelled in palisander, rosewood and African wengé. Throughout the first class accommodation, the carpeting was natural pale grey wool, upon which furniture in tomato red, orange and bottle green was placed. The first class restaurant, a particularly elegant space, featured a large abstract tapestry panel by Urup Jensen in collaboration with Astrid Kahn and a bronze bust of Sir Winston Churchill by Oscar Nemon. This was described in the ship's inauguration brochure as being 'characteristic and forceful'. Aft of the galley, its second class equivalent, in turquoise and blue shades, had an abstract photo mural of ripples in water by Keld Helmer Petersen. Further aft was the second class saloon in mustard and brown tones with a full-width curving glazed bulkhead to give a fine view of the ship's wake. One deck below was the nightclub, located to the stern and well away from cabins so as not to disturb anyone. It was panelled in teak with bright orange and red colours, a copper and aluminium dance floor and a cocktail bar in front of another glazed rear bulkhead overlooking over the stern. Throughout, there were glazed partitions to allow light to flood through the ship by day and full length curtains, which could be drawn at night, to make the same spaces cosy and intimate.
From 1967 until 1974, the *England* and *Winston Churchill* sailed mostly on the

A fine view of the pristine second class restaurant on the *Winston Churchill*. As with the equivalent space on *England* (see above), this too was later converted to a cafeteria. (Kay Kørbing)

The *Winston Churchill*'s second class saloon, located aft, had a fine spread of windows overlooking the stern. The seats in the centre were mustard coloured, whilst those around the perimeter were in brown. (Author's collection)

Esbjerg-Harwich route, but starting in the winter of 1966, the *England* went on cruises to the West Indies and West Africa. The vessel was fitted with an outdoor swimming pool and carried fewer passengers than on North Sea service in one class only. From 1970, DFDS abandoned the class distinctions on all its ships and the facilities became available to everyone.

The *Winston Churchill* continued in service between Esbjerg, Newcastle, the Faroe Islands and Gothenborg. While sailing from there to Newcastle in August 1979, the ship was nearly lost when she ran aground on Vinga Island and the engine room flooded. Everyone was rescued and the vessel was salvaged. It was not until the following summer that repairs, which included the manufacture and fitting of a new bottom, were completed. Later, in 1987, the *Winston Churchill* commenced a series of highly popular spring and autumn cruises to the Norwegian fjords and Baltic ports. After a sensitive refurbishment at Rendsburg in 1989, she remained in DFDS service until 1996 when she suffered damage caused by an engine room fire while bunkering at Esbjerg. Sold to the Florida-based Norwegian shipowner, Normann Tandberg, for service in the Gulf of Mexico as the *Mayan Empress*, she was sent to the Westcon shipyard at Ølensvåg near Stavanger for repair at Esbjerg. Sold to the Florida-based Norwegian shipowner, Normann Tandberg, for service in the Gulf of Mexico as the *Mayan Empress*, she was sent to the Westcon shipyard at Ølensvåg near Stavanger for repair, but unfortunately she became embroiled in a financial dispute between the yard and the owner when allegedly he did not pay for the work carried out. Eventually, after a legal settlement, the ship was ceded to Westcon in lieu of payment. There she remained slowly decaying until December 2003, when she was sold to the Dubai-based Veesham Shipping interests and hoisted the North Korean flag. Given her age and worn condition, however, she is unlikely to see further service.

DANISH SHIP DESIGN 1936-1991

MS *PRINSESSE MARGRETHE* (II) AND MS *KONG OLAV V* (II)

Dressed overall, the sparkling new *Kong Olav V* touches the water for the first time. (Author's collection)

While the finishing touches were being added to the *Winston Churchill* at Riva Trigoso, two new ships for the Copenhagen-Oslo route – the second *Kong Olav V* and *Prinsesse Margrethe* – were also under construction for delivery in 1968 and a further broadly similar pair for the Copenhagen-Aalborg route were to follow in 1969. The new 'Oslo Boats' were even more luxurious than their predecessors. With much-needed full-length car decks and more substantial superstructures, they were perhaps slightly too short and tall to be truly elegant, but within they were generously appointed.

Kay Kørbing was then asked to design the interiors for these as well, and so he devised vibrant schemes in turquoise, blue, orange and red tones. The first class accommodation was divided into a number of intimate rooms with glass partitions. The centrepiece was the nightclub with a brass-panelled circular bar and dance floor. In the hallway there was a colourful mural by Per Arnoldi to the rear of the staircase. New innovations for DFDS were the Pop Room, a disco that was located low down in the hull where it would not disturb other passengers, and a hairdressing salon. As the Oslo service took 16 hours, these ships also had extensive restaurants, bars and shops. By this stage, the lure of tax-free shopping was becoming an ever-more lucrative enterprise for Scandinavian ferry operators.

The *Prinsesse Margrethe* and *Kong Olav V* were subjected to monstrously ugly rebuilds to increase their capacities at Aalborg Værft in the mid-1970s. These involved extending their superstructures aft to make more cabins and heightening their funnels with new lattice-work structures to keep smoke clear of the enlarged sun decks. Replaced by larger vessels in 1983, they were sold to Chinese interests. The former *Kong Olav V* eventually became a Singapore-based casino ship called the *New Orient Princess*, but was destroyed by fire in 1993. The *Prinsesse Margrethe* now sails in the Eastern Mediterranean as the Cypriot cruise ship *Princesa Cypria*.

The second *Prinsesse Margrethe* makes progress at Riva Trigoso while her sister ship, *Kong Olav V*, is almost ready for launching in this view, taken from one of the shipyard's cranes. (Author's collection)

The *Kong Olav V* in the new DFDS Seaways white livery, introduced in 1972. At this point, the ship was refurbished and a new sun lounge was added above the bridge. The existing space, aft, became a discotheque called the Seahorse Club. Forward on the saloon deck, the partition between the smoking saloon and the former-first class nightclub was removed and a single larger space was created, known as the Saga Lounge. (Author's collection)

DANISH SHIP DESIGN 1936-1991

The *Kong Olav V*'s smoking salon, located forward on the saloon deck, was panelled in teak with brass inlays between the panels and turquoise upholstered seats. The Lyfa ceiling lamps, also designed by Kay Kørbing, have matching brass shades. (Author's collection)

The first class stairwell on the *Kong Olav V* appeared to float in front of a large psychedelic mural panel by Per Arnoldi, featuring rainbows, clouds and butterflies. (Author's collection)

MS *AALBORGHUS* AND MS *TREKRONER*

The *Trekroner* on sea trials in the Gulf of Genoa in April 1970, before commencing a short and unsuccessful 6-month stint on the Copenhagen-Aalborg route. (Author's collection)

In contrast, the Copenhagen-Aalborg service was a short overnight hop linking the north of Jutland with the Danish capital. It was very popular in the 1960s as passengers could board the ships late in the evening and arrive the next day early in the morning for a full day's business in the city.

DFDS's new car ferries for this route may have been similar in outward appearance to its Copenhagen-Oslo ships, but they were actually quite different in terms of layout and facilities. For starters, they had bow and stern (as well as side) car deck doors. As with their Copenhagen-Oslo sisters, they had a series of saloons with aircraft-style reclining chairs for second class passengers on each side of their car decks. As DFDS's domestic routes carried many unberthed 'deck class' passengers, this development enabled all passengers at least to be guaranteed a comfortable seat for the duration of the crossing.

The greater parts of their superstructures were given over to cabins and public rooms were located in the after portions of their promenade and upper decks. First class facilities consisted of a cosy lounge forward of amidships with a small restaurant on either beam. Their spacious cafeterias were located aft with a small, robustly appointed second class lounge and bar on the promenade deck below.

The development of Denmark's motorway system and shorter ferry links further south, however, meant that by the time the first of its new ships, the *Aalborghus* was delivered, the route was less busy. A strike at the shipyard delayed the completion of the *Trekroner*, the sister ship, until 1970.[35] By then, DFDS had decided to abandon its domestic routes as it could not compete with the state-subsidised railway-owned DSB ferries; worse still the Swedish authorities also prevented the ships from calling at Helsingborg en-route to enable tax-free shopping

DANISH SHIP DESIGN 1936-1991

The *Kong Olav V's* first class nightclub was decorated in red and gold and had a circular cocktail bar and metal dance floor as its centrepiece. This 1960s DFDS publicity photograph hints at the glamour and elegance of the company's Oslo ferries. (Author's collection)

onboard. The outcome was that the route was abandoned and a search for a new use for its ships began.

DFDS had observed that tens of thousands of Germans and Scandinavians headed south towards the Mediterranean in their cars each summer, yet the local shipping companies had very few ships with car carrying capacity. Spotting what appeared to be a good business opportunity, DFDS had the two ships transformed from overnight ferries into sleek, white ferry-cum-cruise liners at the Societé de Anciens Établissement Groignard shipyard in Marseilles. Kay Kørbing remembers the conversion:

'As the *Aalborghus* and *Trekroner* had been designed for a route on which they sailed at 10 pm and arrived at six the next morning, they had very few public rooms – a cafeteria and a couple of bars and that was it. Also, they had little deck space, so we dismantled a whole area of cabins to make a big lounge on the saloon deck and more to the stern to make a nightclub. We made reclining chair lounges into cabins with private facilities, so we were able to re-use a lot of materials. One of the sun decks was extended out to the stern to make space for an outdoor swimming pool with glass shelter screens. All this was done simultaneously on both ships when they were in drydock, one behind the other.'[36]

The ships were transformed from black-hulled ferries into the gleaming white *Dana Corona* and *Dana Sirena*. The new livery, nomenclature and marketing name 'DFDS Seaways', painted on the hull in elongated blue letters, soon spread to all of DFDS's passenger ships.

The two ships then sailed for Genoa, where they tied up on either side of the terminal at Port Calvi. With streamers flying and sirens blowing, they sailed on 25th June 1971, the *Dana Corona* bound for Tunis and Alicante and the *Dana Sirena* for Malaga and Palma de Mallorca. The route was a great success at first, but gradually locally-

As with the recently-introduced Copenhagen-Oslo ships, the *Aalborghus* and *Trekroner*'s first class lounges were located amidships with restaurant facilities on either side. To create an intimate atmosphere, there were brass hanging lamps over the tables. As the ships sailed late and arrived early, there were no entertainment facilities, however. (Kay Kørbing collection)

The second class bar on the *Aalborghus* had a teak floor like an outdoor sun deck and gold mosaic panelling behind the bar. With turquoise upholstery, it was bright and hard-wearing. (Kay Kørbing collection)

DANISH SHIP DESIGN 1936-1991

The former *Aalborghus* and *Trekroner* lie one behind the other in the drydock at Marseilles during the Spring of 1971. Note that larger windows have been cut into the superstructure as cabins have been removed to create a large lounge and bar space. (Author's collection)

With its transformation complete, the gleaming white *Dana Corona* (ex *Trekroner*) leaves Genoa en-route to Tunis and Alicante. Note how the decks have been extended aft to create an outdoor lido area. (Author's collection)

The *Dana Sirena* is seen sailing through a calm Mediterranean during her first season after rebuilding. The all-white DFDS Seaways livery emphasises her rather short hull and tall superstructure. (Author's collection)

In what was originally a block of first class cabins, a large saloon called the Hamlet Lounge was created. This was similar in style to these popular and relaxing spaces on DFDS's North Sea and Copenhagen-Oslo ships. (Author's collection)

owned, state-subsidised companies acquired their own car ferries and the Danish invaders could not compete.

With a downturn in the late 1970s, the *Dana Sirena* was withdrawn from Mediterranean service and then led a nomadic life being chartered out to other shipping lines, mainly in Northern Europe. In the autumn of 1979, she was rebuilt by Aalborg Værft with a new, tall funnel and extended passenger accommodation for a return to Mediterranean service. At this point, the two ships swapped names and the unrebuilt *Dana Sirena* (ex *Dana Corona*) was tried on a new service from Genoa to Patras, Heraklion and Alexandria, while the improved *Dana Corona* (ex *Dana Sirena*) returned to the Western Mediterranean. Unfortunately, the financial results were still poor, and so the routes were abandoned in November 1982 and the ships sold.

The *Dana Sirena* followed a number of other DFDS passenger ships to the Red Sea, sailing mainly from Suez to Jeddah carrying huge numbers of pilgrims and migrant workers, mostly unberthed and sleeping on deck – a service far removed from the luxurious ship she had been under DFDS ownership. Re-named *Al Qamar El Saudi* (the Saudi Moon) and looking the worse for wear, she featured in the 1988 Michael Palin TV series 'Around the World in 80 Days.' The ship subsequently became *Al Qamar El Saudi El Misri*, owned by Egyptians. Unfortunately, she suffered a massive boiler explosion and was gutted by fire off Sotage in May 1994. An American destroyer rescued 110 pilgrims, returning from the Haj, and the former *Dana Sirena* sank. The *Dana Corona*, meanwhile, was sold to the Chinese in 1985 and still exists as the Dalian Steamship Company's *Tian E*.

Passengers relax on the upper sun deck of the *Dana Sirena*. To the rear is the semi-outdoor Viking Bar (Author's collection)

MS *VISTAFJORD*

Similar in appearance to her older sister, *Sagafjord* (though taller by one deck), the *Vistafjord* was the last of the Norwegian America liners. She is seen here approaching Tilbury in 1979. (Ambrose Greenway)

By the 1970s, the viability of Atlantic crossings was in terminal decline and Norwegian America Line wisely decided to concentrate on the provision of up-market cruises. In December 1969 an order was placed with Swan Hunter Ltd on the Tyne for the construction of a further luxury cruise vessel, designed to the same high standard as the *Sagafjord*.

The new ship was launched on 15th May 1972 as the *Vistafjord* and was handed over to her owner exactly a year later. To continue the standard set by the *Sagafjord*, a great deal of advanced thinking went into the design, yet the external appearance retained the same attractive lines that characterised all of the company's post-war ships. The superstructure was an extra deck taller but it had a slightly more severe profile without the inward curves of the earlier ship's bridge and funnel structures.

The *Vistafjord* had a gross tonnage of 24,291 grt and could carry 550 passengers in one class served by 390 crew. Two Sulzer diesel engines gave a maximum speed of 22.5 knots. The extra deck reflected the ship's larger cruising capacity and when new, it was claimed that no ship in the world offered so much space per passenger. Most of the cabins were outside rooms and there were 164 single cabins which could be inter-connected to become suites. The ever popular Garden Lounge, forward hallway, Norse Bar and Library were again designed by Kay Kørbing in a similar style to the same facilities on *Sagafjord*. Although the hallways were panelled in teak, overall the *Vistafjord* used less woodwork as the designers anticipated the forthcoming SOLAS 1974 fire safety regulations for passenger vessels.[37] As with the *Sagafjord*, the *Vistafjord* was adorned with a significant number of original pieces by noted Norwegian and Danish artists. A tapestry in the main entrance hall on Veranda Deck was woven by Anne Lise Knudtzon, based on a cartoon by Knut Rumohr. Lise Honore produced a series of abstract white clay sculptures in low relief which were hung against the dark blue walls of the Norse Bar while Carl B. Gunnarson designed a wrought iron and brass decorative panel on a mirrored background, installed behind the bar in the Club Viking. Tove Krafft painted a mural in the main staircase between the Promenade Deck and Veranda Deck and the Vista Dining Room contained several lithographies by Suzanne Ogaard and a large brass decoration in the room's centre by Jorlief Uthaug. The same artist also sculpted a larger brass and enamel decoration for the bulkhead in the adjacent hallway, which could be viewed from the dining room through a plate glass partition.[38] The last mentioned remain in situ today, although many of the other artworks have since been removed by the *Vistafjord*'s subsequent owner, Cunard.

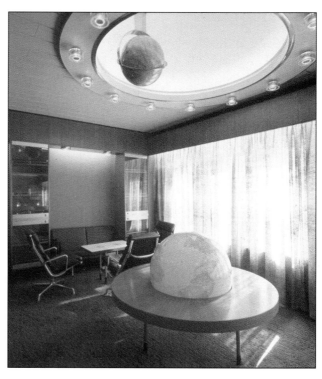

The *Vistafjord*'s library and writing room featured an illuminated globe as its centrepiece. (Kay Kørbing collection)

The *Vistafjord* sailed on her maiden voyage from Oslo to New York on 22nd May 1973 and has since been employed cruising on both sides of the Atlantic. In cruise service, the *Sagafjord* and *Vistafjord* rarely met as the former tended to be American-based, cruising in the Pacific to the Far East or on summer voyages to Alaska. The *Vistafjord*, meanwhile, developed a peerless reputation for her European-based cruises, except in winter when sailing on Caribbean itineraries. Sadly, the *Vistafjord* was the last Norwegian America passenger liner to be built. Apart from the huge operating expenses of a ship with such a high crew-to-passenger ratio, escalating building costs made it uneconomical for such a liner to be constructed in the future.

The *Vistafjord* was sold to Cunard along with the

The Garden Lounge on the *Vistafjord*. (Kay Kørbing collection)

Furniture in the *Vistafjord*'s card room and reading lounge: throughout, the ship's table tops featured inlaid ceramic tiles by a variety of Norwegian artists. (Kay Kørbing collection)

Sagafjord (see above). In 1995, she acquired extra superstructure containing penthouse cabins which obscured the base of her funnel. In 1998, Cunard was finally sold by Kvaerner to the Carnival Corporation, the world's biggest cruise line and a company anxious to expand further into the deluxe end of the cruise market. Carnival announced plans to consolidate Cunard as a recognisably British brand name and to emphasise the company's history and traditions. Wearing full Cunard livery for the first time, the former *Vistafjord* was renamed *Caronia* in Liverpool in November 1999. Inside, the ship is now barely recognisable as the modern Scandinavian liner she once was, indeed the new décor is a pastiche of earlier styles of shipboard design. Notwithstanding this impairment, the *Caronia* remains a popular cruise liner and is much admired for her luxury worldwide itineraries. She has recently been sold to Saga and will re-join her former Norwegian America fleetmate, *Saga Rose*, in the Autumn of 2004.

The *Caronia* (ex *Vistafjord*) leaves Rosyth on the River Forth during a round Britain cruise in Summer 2000. The photograph is taken from the Forth Road Bridge. (Author)

DANISH SHIP DESIGN 1936-1991

MS *DANA REGINA*

The handsome *Dana Regina* is seen at Esbjerg from the stern of the *England* in 1978 during the heyday of DFDS's North Sea ferry services. (John Peter)

By the early 1970s, growth on DFDS's long-established Esbjerg-Harwich route was so great that the company required a further large passenger liner to replace the ten-year old *England*. This time, the order went to a Danish shipyard, the Aalborg Værft, and, as the hull was too long for the slipway, she was launched in August 1973 without her bow section, which was added later in dry dock. The cost of building ships had risen steeply since the *Winston Churchill* was built, and at 101 million kroner, the new vessel, code-named *Dana Futura*, cost twice as much. Kay Kørbing was also responsible for the interior design of this ship:

'When my office was developing the so-called *Dana Futura* project, we were also working on parts of the *Vistafjord* for Norwegian America Line and, although they were designed for very different purposes, some of the ideas from *Vistafjord* were incorporated in the design. The most obvious influence was in the external profile, which I helped to style in conjunction with DFDS's technical department. I think that is where the tall, tapering funnel and the overall shape came from. The new Royal Viking Line cruise ships were another influence, as were a number of recent ferries and cruise liners designed by the Knud E. Hansen company, at least for the layout of the public rooms and in the use of large picture windows.[39] There were also the new SOLAS fire regulations to consider which prevented us from using wood panelling. Consequently, the *Dana Regina*, as she was eventually named, used a lot of moulded fibreglass and other modern finishes, so she appeared very bright and spacious.'[40]

The 12,192-ton *Dana Regina* was Denmark's largest passenger ship since the *Frederik VIII* of 1913. As such, the vessel was a showcase for the newest and best in Danish design and technology. On the technical side, design work was carried out by Bryan Corner-Walker and his team, who worked in close co-operation with the shipyard and with Kay Kørbing. Corner-Walker recalls that 'The Aalborg shipyard was a most efficient enterprise with a large, effective drawing office and great

The lounge adjacent to the Codan Restaurant was the perfect spot in which to enjoy a pre-dinner drink. Note too the splendid open-tread spiral staircase, accessing the Mermaid Bar above, to the rear of this view. (Kay Kørbing collection)

engineering expertise. Certainly, it was a great pleasure to work with them as their workmanship was first class throughout and that included the interior outfitting.'[41]

Powered by B&W diesel engines and designed for a 21.5 knot service speed, the sleek hull had a bulbous bow. The new vessel could carry 975 passengers and 250 cars and the bow car deck doors were of a novel 'clam shell' design, which opened out and to the sides of the hull (a primitive hinged version of this type of door had first appeared on a series of ferries designed and built by Wärtsilä in the mid-1960s, such as the *Finnhansa*, *Finnpartner* and *Finlandia*). One reason was that during a storm in 1968, mountainous seas had smashed open the *Winston Churchill*'s lifting bow visor and only the brave action of the crew in tying hawsers round the locking

The *Dana Regina*'s Codan Restaurant was spacious and stylishly furnished. As with previous Kay Kørbing designs, the ship's structural members were expressed in the ceiling. (Author's collection)

DANISH SHIP DESIGN 1936-1991

The Mermaid Lounge was spacious and every bit as elegant as similar spaces on contemporary Scandinavian-owned cruise ships. Kay Kørbing's KK39 lounge chairs are mounted on pedestals – a bespoke design unique to the *Dana Regina*. (Author's collection)

pins had prevented the ship from being lost. With the new design, such an accident would be impossible. Furthermore, the *Dana Regina* was fitted from the outset with a biological sewerage disposal system, an electric waste separator for the kitchens and an engine room incinerator to deal with waste oil and solids, making her one of the most environmentally-friendly ships of her generation.

Inside, as all the public rooms served one class of passengers, Kay Kørbing designed a spacious arcade with lounge chairs and floor-to-ceiling windows along the port side with kitchens and services to starboard. A cosy bar, called the Admiral's Pub, shops, a playroom and conference rooms were located along the arcade. At the forward end of the saloon deck was the Codan Restaurant, containing a mural of the phases of the Sun and Moon by Erik Clausen which matched the blue and gold colour scheme. This was connected to the circular Mermaid Bar and adjoining Bellevue Lounge, both on the deck above, by an elegant, open tread staircase with smoke glass balustrades. Above this, there was a golden ceiling dome inset with twinkling lights. The colour scheme was lilac, red and gold and there was an abstract tapestry by Margrethe Agger. Towards the stern of the Saloon Deck was the Scandia Coffee Shop with its cheerful wavy copper ceiling and leather-upholstered aluminium chairs and a mural by Rolf Middelboe on the forward bulkhead, which concealed the servery. The Compass Club, the ship's nightclub, at the stern had floor-to-ceiling windows round three sides, a back-lit ceiling with oblong slits, similar to the Queen's Room in the *Queen Elizabeth 2*, and low-slung red stools and chairs by Jan Ekselius. One deck below amidships was the entrance hall with painted timber relief panels on the bulkhead by Markan Christensen. The main staircase, adjacent, featured a circus scene by Per Arnoldi. Outside, there were wide, sweeping sun decks and the lifeboat davits were mounted above the boat deck, as on the *Vistafjord*, to give uncluttered expanses of teak on which the passengers could promenade. With so many specially commissioned artworks and with such careful attention to every design detail, the *Dana Regina* was the height of 1970s stylishness and it was rumoured that, as well as sailing between Harwich and Esbjerg, she might also go cruising.

The bright and airy cafeteria with its wavy copper ceiling and brown leather chairs in brushed aluminium frames was located at the after end of the ship's portside arcade. The servery was concealed by a pointalist mural by Rolf Middelboe and the aluminium-framed chairs were designed by Kørbing and manufactured by Poul Cadovius. (Author's collection)

Such luxury, however, came at a price. Bryan Corner-Walker points out that 'Kørbing was a very good architect indeed, but he was also quite astronomically expensive as all of the materials he specified had to be the best imaginable – crystal lights where glass would have sufficed, leather, plate glass and specially commissioned artworks. Unfortunately, as the 1970s progressed, there was a growing feeling amongst DFDS's directorate that, however beautiful, such expense could not be justified from a business viewpoint.'[42] This was most unfortunate as there is little doubt that the interior designs of DFDS ships until this point made a deep impression on their passengers and, furthermore, acted as floating ambassadors for all that was good about Danish design, cuisine and hospitality. What is more, they were robust and stood the test of time remarkably well.

The newcomer was named in Copenhagen by Queen Margrethe II of Denmark on 1st July 1974 and, after a visit to the Pool of London (when she became the largest ship ever to sail through Tower Bridge) she entered the Esbjerg-Harwich service on 8th July alongside the *Winston Churchill*. The older *England* was then cascaded to the seasonal Newcastle-Esbjerg route. The *Dana Regina* was to be Kay Kørbing's last ship design for DFDS as a new technical director, who superseded Bryan Corner-Walker, evidently did not appreciate his work. Indeed, during the ensuing years, many of Kørbing's designs for the company's ships were unfortunately altered to their detriment (see below).

The *Dana Regina* was transferred to the Copenhagen-Oslo route in late-1983 and sold five years later to a Florida-based company, Marne Investments for use as a casino ship – possibly in the SeaEscape fleet. The ship was re-registered in the Bahamas but, before delivery, she was re-sold to the Swedish firm Norström and Thulin for service in the Baltic. Known at different times as *Nord Estonia* and *Thor Heyerdahl*, she is now the *Vana Tallinn*, operating between Kapellskär and Tallinn for the Estonian Tallink and painted in a garish colour scheme of blue and red stripes.

MS *TOR BRITANNIA* AND MS *TOR SCANDINAVIA*

The long, sleek *Tor Britannia* and *Tor Scandinavia* were two of the North Sea's most outstanding passenger vessels and are seen near Amsterdam in their original Tor Line livery. (DFDS)

As the *Dana Regina* had received considerable positive publicity in the shipping press, it was not surprising that the Swedish-owned Tor Line, part of the Salén Group, requested Kay Kørbing's assistance to help design two 15,650-ton cruise ferries for its triangular Immingham and Felixtowe to Gothenburg and Amsterdam routes. Ordered from the Flender Werke shipyard at Lübeck, the *Tor Britannia* and *Tor Scandinavia* were to be delivered in 1975 and 1976 respectively.[43] In order to fit through the tidal lock at Immingham, the ships' width was restricted to 23.6 metres, and so to compensate they were exceptionally long. To cut sailing times significantly, they were also very fast with four powerful Pielstick PC3 engines giving a top speed of over 26 knots. With sharp clipper bows, long and low superstructures, massive streamlined funnels and a sweeping tier of sun decks at the stern, the 'Tor' ships were impressively good looking. Thomas Wigforss was appointed as Project Manager shortly after the building contract was signed with the shipyard and he recalls the background to the ships' design and construction:

'Tor Line began in 1966 with two ships, running in competition with the Swedish Lloyd, for whom I then worked. Swedish Lloyd was part of a consortium called the England-Sweden Line which actually consisted of three partners – Rederi A/B Svea of Stockholm, Ellerman's Wilson Line of Hull and Swedish Lloyd itself. These companies each built a new ferry for their North Sea routes, but these were fairly slow and, even when new, rather traditional both in design and routing. The Swedish Lloyd ship, *Saga*, for example ran from Tilbury to Gothenburg – a very lengthy route that was slowed even more by having to navigate up the Thames, yet this ship could only manage about 18 knots. The first Tor Line ships ran on shorter routes from Immingham and Felixtowe and could do over 21 knots and Tor Line's aim was to shorten its service to Sweden so that a single crossing could be completed in under 24 hours. That is why the new *Tor Britannia* and *Tor Scandinavia* had such powerful machinery.'[44]

The *Tor Britannia*'s restaurant, aft on the saloon deck, was brightly coloured and featured troughs of flowers around its perimeter with a series of sculptures of dancers. Conceptually, it was similar to the earlier Garden Lounges on *Sagafjord* and *Vistafjord* with a circular design focusing attention on the middle of the room at night and large picture windows to give good sea views by day. (Andres Bergenek collection)

As to their design and construction, 'in the 1970s, Tor Line was largely owned by the Swedish Salén Group with a minority shareholding belonging to the Dutch firm KNSM. Knud E. Hansen A/S carried out the initial design work for the ships, but Salén's own extensive technical department then took over.'[45]

At that time, a young Swede by name of Klas Brogren and his Danish school friend,

A part of the *Tor Britannia*'s cafeteria amidships, showing a section of the long mural of brich trees by Lisa Grönvall and Cliff Holden which wrapped around three sides of the ship's galley. Unfortunately, this was removed when the space was converted to an à la carte restaurant by DFDS in 1991. (Klas Brogren)

The arcade on the *Tor Britannia*: Not only did this connect the main stairwells and public rooms on the ship's restaurant deck, but it also was a most pleasant space in which to relax and watch the passing seascape. Bruno Mathsson – one of Sweden's most distinguished modern furniture designers – produced the chairs and the lighting was by Børbing himself. (courtesy of Klas Brogren)

Ivar Moltke, were teenage ship enthusiasts. Brogren recalls that, just for fun, they made drawings for a speculative 'ferry of the future', which they sent to Tor Line's passenger manager, Lars Wikander. Evidently, he was both intrigued and impressed as the two youths were invited to join Tor Line's design team to provide fresh ideas and to sample, document and report upon ferries belonging to Tor Line's competitors. He recalls that:

'For us, this was a dream come true and most exciting. Thinking back, it also reflected the fresh and innovative thinking of Tor Line's management. The company wanted modern, colourful ships in tune with the best design standards of the time and so it was decided that, because of his fine reputation, Kay Børbing should be asked to draw up the majority of the interior spaces. His work was perceived to be very elegant, but possibly a little too understated and that is why Tor Line also engaged the highly regarded Finnish designer Vuokko Laakso to add rich and exuberant colours to the ships' entertainment spaces. We wanted the elegance of Børbing on the restaurant deck to contrast with the colours of Laakso on the saloon deck.'[46]

Børbing, however, had already approached Tor Line with a view to being awarded a share of the design work on the new ships:

'When my relationship with DFDS ended abruptly upon the completion of *Dana Regina*, I was very anxious to find more work of a similar kind – and quickly – to keep my staff fully employed. I had read in the newspapers about Tor Line's plans and so I contacted them at once. This was the only occasion when I had actually to contact a potential client, rather than them approaching me. Happily, I got a positive response but, as there was no further correspondence for a long time, I wrote again to say that if the interiors of these ships were to be designed properly within the required timeframe, then work had better begin within the next couple of weeks. Only a day or so later, I was telephoned and asked to begin designing. Thus began a very constructive relationship with Thomas Wigforss and the other Tor Line people. This resulted in two very fine ships being constructed – ships that moved the ferry design evolution some way forward.'[47]

Kørbing, infact, rose to the occasion magnificently and, as well as contributing his own work to the cabins, corridors, hallways and catering facilities, he was also required to co-ordinate the work of the other artists and designers, recruited from Britain and Scandinavia, who had been assembled by Tor Line. The overall concept involved vertically dividing the ships' accommodations into three colour-coded sections to make passenger orientation as straightforward as possible. Thus, the forward third of the passenger decks and the forward stairwell were finished in shades of blue, with the midships section in predominantly orange tones, while the aft section was in green. A precedent for this was the Cunard liner *Queen Elizabeth 2*, which Kørbing had visited during final outfitting at Clydebank in 1968 – this was probably the first passenger liner to have colour-coded staircases. The detailing of the cabin corridors was also similar to those onboard the 'QE2' as the doors were all recessed in pairs with diffused lighting above to illuminate each niche. The majority of cabins were located forward, away from the noise of the engine uptakes, and the public rooms were spread over two decks towards the stern. The main deck contained a large cafeteria and grill forward and this was linked to the restaurant by an attractive arcade on the port side. These were decorated in the vivid colours typical of the period.

The restaurant, located aft, had a layout similar to the Garden Lounges on the *Sagafjord* and *Vistafjord* with large windows on three sides giving panoramic sea views. A ceiling dome with concealed perimeter lighting ingeniously gave the illusion of extra height and helped to define the central seating area. On *Tor Britannia*, four raised flower troughs split the space into a number of smaller sections and contained bronze statuettes by Axel Olsson of young people dancing, while on *Tor Scandinavia* the troughs contained Örrefors crystal sculptures made by Carl Fagerlund. Indeed, the extensive art installations were outstanding features of the two ships which the shipowner Christer Salén had specially commissioned. Both restaurant and cafeteria on *Tor Britannia* featured wrap-around murals by Lisa Grönvall and Cliff Holden, depicting seagulls over waves and birch trees respectively, while *Tor Scandinavia*'s cafeteria featured murals with 'bowler hat' motifs by Per Arnoldi. On *Tor Britannia*, the port side arcade, which connected the food outlets, was adorned with decorative panels by Gösta Werner made of colourful spinnaker sailcloth, while there were paintings by Lennart Jirlow on *Tor Scandinavia*. Ernst Grundtvig, meanwhile, decorated the stairways with abstract compositions made of timber rescued from old sailing ships, while Ulf Törneman made iron bas-reliefs of dancing teenagers for the *Tor Scandinavia*.

On the boat deck, Vuokko Laakso boldly decorated the ship's entertainment spaces. Amidships, there was a large circular nightclub in orange and red tones with slightly terraced seating. The casino was amidships and the discotheque, decorated in orange and green, were located fully aft where they would be least disturbing. These spaces were connected by a narrow arcade in which there was a most imaginative children's play area, featuring a sailing ship on which to play. Kay Kørbing, however, retained overall control of the entire project to ensure a high degree of visual coherence. According to Thomas Wigforss:

'Kay Kørbing was most concerned with the detailed design of the ships. He went to Norway with me to supervise the weaving of samples for the carpets, then to Germany to visit the supplier of anodised aluminium – a material he loved – which was to be used for stair handrails, door frames and other details. When at the shipyard, he would produce small pieces of brushed aluminium from his pocket to check that what was being used was exactly to specification. There, mock-ups were constructed of the different cabin types and members of Tor Line staff were brought down from Sweden to sleep in them to check that everything was as ergonomically designed as possible. Consequently, many evenings were then spent in our hotel refining the design details, with Kørbing sketching ideas of how he felt that certain aspects should be carried out. For me, this process was fascinating and most enjoyable as he was a most pleasant gentleman to work with. At the shipyard, however, he was utterly resolute and insisted that the work should be done exactly as he intended. The trouble was that the Flender yard was not the most up-to-date and neither its management nor the workers had much experience in fitting out large luxury passenger ships of this kind, so they often took a great deal of persuading – but somehow Kørbing managed to get his way.'[48]

The crew's accommodation, designed by the Swedish architect Hans Nilsson, brought about a transformation in working conditions on North Sea passenger ships – it being every bit as spacious and luxurious as the passengers' facilities. For example, the mess and the lounge for officers and crew were placed on the same decks as the passenger's restaurants and lounges. Almost all 143 crewmembers had individual outside cabins placed on the ships' top decks 8 and 9 with a few extra cabins found aft on Deck 5. Ordinary crewmembers each had a single cabin, but with a shared toilet and shower units. Officer's cabins were fitted with individual private facilities. (In older vessels, most crewmembers shared 2-berth cabins below the car deck. Such areas, often located on top of the engine room, were noisy and often too warm and many cabins had no access to natural daylight.)

Christer Salén explained that, as the crew would spend 14 days aboard on each shift while the passengers would spend only 24 hours, naturally they should be entitled to the best areas in the ships. There were several reasons for this new way of thinking. The awareness of the importance of a pleasant working environment to increase both loyalty and productivity had risen in Scandinavia. In the mid-1970s, Swedish unemployment rates were extremely low and the maritime trade unions powerful. Consequently, Tor Line had had problems in hiring and retaining werestaff on its existing vessels, especially in the restaurant and hotel departments. Together, these factors stimulated a desire to make *Tor Britannia* and *Tor Scandinavia* the most luxurious and democratic ships to live and work in for officers and ordinary crewmembers alike.

Kay Kørbing also became involved in the exterior styling of the two ships – in particular their tiered sun decks, located aft. Thomas Wigforss devised their distinctive funnel design, but Kørbing took a great deal of trouble to ensure that the overall effect was powerful but harmonious with long straight lines and sweeping curves. He also suggested the application of the blue and white Tor Line livery in such a manner as to emphasise their great length and speed.

Even although first and foremost intended for North Sea routes, the 'Tor' ships featured outdoor lido areas with sheltering glass screens on their topmost decks forward of the funnel – facilities that were rarely used in daily service but which were, according to Thomas Wigforss, an invaluable publicity tool in Tor Line's

The *Tor Britannia*'s nightclub was designed by Vuokko Laakso using ceiling finishes, plate glass doors and other finishes by Kørbing, who had overall control of the ship's internal appearance. The 'Tor' sisters and the *Dana Regina* were amongst the first passenger vessels to make use of Dampa metal ceilings throughout their passenger accommodations. These high-performance finnishes were not only fireproof, acoustically-absorbant and lightweight, but also enabled easier access to service the ducting and pipework above. (Klas Brogren)

marketing material. In the Tor Line era, only the *Tor Scandinavia* ever left the North Sea to sail in sunnier climes when she was also chartered during successive winters to sail to the Middle and Far East to promote Dutch goods and trade. The last such charter was during the 1982-83 winter. It was on their regular North Sea service, however, that the 'Tor' sisters proved to be outstanding successes and they were subsequently used on a variety of routes linking Britain, Denmark, Sweden, and The Netherlands.

When sold to DFDS in 1982, the Salén group had most of their artworks removed and these are now on display in that company's offices in London and Gothenburg. Only the decorative panels in their aft stairwells remained in situ. Otherwise, the ships remained largely intact until 1991 when they were sent to Blohm and Voss in Hamburg for a major rebuild. During this work, the Swedish architect Robert Tillberg produced new interiors which almost obliterated the original designs to the extent that only their buffet restaurants remained recognisably intact. They were re-named *Prince of Scandinavia* (the former *Tor Britannia*) and *Princess of Scandinavia* (*Tor Scandinavia*). In 1996, the *Prince of Scandinavia* was chartered to the Tunisian company Cotunav to sail in the Mediterranean, but returned to the North Sea the following season. After further re-building work at the Gdansk shipyard in Poland in 1998, during which time they were fitted with sponsons to comply with new SOLAS regulations, 1998, during which time they were fitted with sponsons to comply with new SOLAS regulations, the ships continued in front line North Sea service. In November 2003, after 28 years of continuous usage, the *Prince of Scandinavia* was sold to Moby Lines for service in the Mediterranean as *Moby Drea* between Livorno in Italy and Olbia in Sardinia.

CHAPTER 3
THE DSB FERRIES

The mid-1970s heralded further major advances in Scandinavian ferry design. As ever more capacity was required for cars and especially freight while foot passengers wanted to shop tax-free, there was a need for larger ferries able to fit existing quays. The popular solution was the 'large block principal', which sought to maximise the space utilisation of a ferry's hull and superstructure. If ferries could not become significantly longer, they would need to become wider and taller. Four ships developed by Stena Line for the charter market in the mid-1970s perhaps best demonstrated the new approach, being essentially shoe box-shaped above the waterline (apart from their bows).[49] While no doubt efficient, profitable and able to offer passengers a far greater range of facilities, the new generation entering service in the early-1980s presented a bluff, hefty appearance.

Kay Kørbing's next interior design projects were all for DSB, the Danish State Railways. Given his belief that ship interiors should reflect their external appearance, he used many innovative design tactics in these ships, exploiting the new spaciousness that the 'large block principle' design method facilitated to best advantage.

In the early-1970s, DSB's management had decided to follow British Rail's example in strengthening its image by instituting a new corporate identity policy. As with British Rail's much-admired scheme, devised by Jock Kinnear and others in Sir Misha Black's Design Research Unit, the new DSB look not only involved the repainting of the trains (using a patriotic livery of red, white and black), new station signs and graphics, but also extended to the actual design of all new trains, stations and ferries. Jens Nielsen was appointed Director of Design at DSB and was responsible for this work. He initiated the process of recruiting architects to work as design consultants and, through his enthusiastic engagement in all aspects of transport design, he built up teams to modernise all departments of DSB.

At that time DSB was Scandinavia's largest ferry operator with a big fleet of both train and car ferries, efficiently linking Denmark's many islands. Principal routes were the Great Belt crossings Nyborg-Korsør and Halskov-Knudshoved, which carried trains and cars respectively, and the Rødby-Puttgarten connection between Lolland and West Germany, run jointly with the West German railway, Deutsche Bundesbahn. While the latter was served by a relatively modern fleet, the Nyborg-Korsør route was seen by DSB as the 'weak link' in Denmark's rail network, served as it was by a motley collection of small ferries mainly dating from the 1950s. At either end of the crossing, Inter-city trains had to be split into several sections to fit onboard. Although DSB had perfected this to a fine art, it was still time-consuming and there was an overriding desire to increase capacity, especially between the principal cities of Copenhagen, Odense and Århus. Furthermore, it was also hoped that the two train and car Great Belt routes would eventually be integrated with up to six large, wide ferries to carry all traffic across.

MS *DRONNING INGRID*, MS *PRINS JOACHIM* AND MS *KRONPRINS FREDERIK*

The first of the three DSB Inter-city ferries displays its rectilinear, but purposeful lines in this publicity photograph, taken shortly after they entered service. (Author's collection)

The initial solution was to order three large train and passenger ferries, each twice as big as its largest predecessor at 11,500 gross tonnes. Each was to have four tracks on the train deck, with the possibility of loading two trains simultaneously through bow doors. From the outset, the ships were designed so that, in future, their superstructures could be raised to incorporate vehicle decks above their train deck spaces. The plan was eventually to build three further ferries, each able to carry trains and road vehicles, to integrate the Halskov-Knudshoved car ferry and Nyborg-Korsør train ferry routes as one.

The architect and industrial designer Niels Kryger, who designed trains and buildings for DSB, prepared the initial design specification and general layout in conjunction with DSB's own technical department. Kryger's most recent ship designs for the Danish government had been two ferries for Bornholmstrafikken, the *Jens Kofoed* and *Povl Anker*. The design of these ships owed much to Kay Kørbing's own work and, indeed, used several of his lighting and furniture designs. As Kryger clearly admired Kørbing, it was perhaps not surprising that at an early stage in the design process in 1977, he was asked to take on a leading role in the design of the passenger spaces on the new ships.[50] This presented a new challenge:
'Naturally, I was very delighted to get this job as these new state-owned ferries, carrying the names of the Danish royal family, were to be at the heart of the country's transport network. Unlike my previous ships, which led comparatively relaxed lives with long crossings and several hours at least in each port, these DSB train ferries were to operate on the most intensive of schedules – crossings of just under one hour with only twenty-five minutes to unload four trains with up to 2,000 passengers and a further twenty-five minutes to load again. All the passengers had to be got from their carriages up to the two main saloon decks to eat, drink and shop, and back down again before the ferry arrived, so that their trains could be offloaded immediately.'[51]

There was close co-operation on all aspects of the design process from the

DANISH SHIP DESIGN 1936-1991

The *Dronning Ingrid* leaves Korsør in a strong breeze in the Summer of 1995. The funnel has been heightened but, otherwise, she is in as-built condition. Constructed for service in relatively sheltered inland waters, one wonders how she will cope with the stormy oceans in her new role as a hospital ship. (Author)

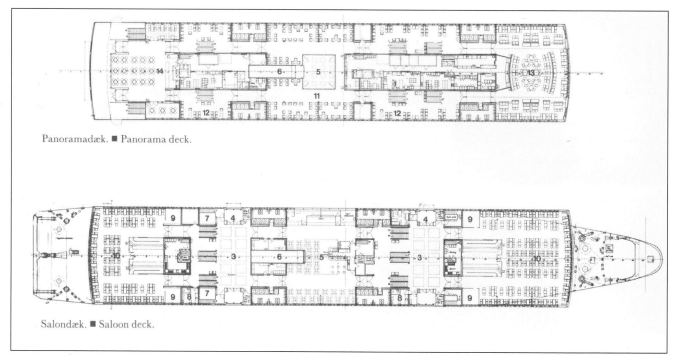

The general arrangement of the two saloon decks of the DSB Inter-city ferries was almost symmetrical with facilities duplicated on each side of the centre line. Note the perimeter circulation layout and the large stair hallways forward and aft of amidships. (Kay Kørbing collection)

beginning, thus enabling the interior design team to influence the ships' structural arrangement. In getting to an appropriate solution, Kørbing stuck to the same rational architectural principals which had informed all of his previous ship designs, but he reached some very innovative solutions:

'To get enough capacity for complete inter-city trains on the train deck, these ferries were to be some 500 feet long and also very broad with four tracks and both central and side trunking running the length of their hulls to access all four tracks on the train deck. Obviously, passenger orientation was going to be a serious issue – how would someone find their place on the correct train again out of maybe fifteen identical carriages parked down there?'

The solution proposed by Kryger was to make an impressive central focus in the middle of each of the ships – the 'ferry square', a two-storey space off which the shops and fast food bars were accessed. The main stair halls were forward and aft of this space, and one was finished entirely in DSB red and the other in blue. The

The ferry square on the *Dronning Ingrid*, showing Arne L. Hansen's wall design on the funnel casing and the snack bar on the opposite bulkhead. A staircase was later inserted to improve circulation as it must have been frustrating for passengers to have had to go forward or aft to get up or down stairs. (Kay Kørbing collection)

stair balustrades, the flooring and even the murals on the walls were made to match. To find their carriage again, passengers only had to remember a track number and the colour of the staircase (incidentally, these were generously proportioned incase the ships were heightened through the insertion of additional car decks in the future). 'The other problem was that once the passengers were on the saloon decks, there was maybe only forty minutes maximum left for them to eat and drink, so the catering facilities would have to be ready and open as soon as the trains were loaded. With the intensive sailing schedule, the only solution was to duplicate all the servery facilities along the centre line of the ship, so that from the passengers' point of view, the starboard side was almost a mirror image of the port side. While the port side serveries were in action, those to starboard could be cleaned, re-stocked and made ready for the return crossing.'[52]

The main restaurants and lounges were located forward and aft to give good views out over the Great Belt. At the forward end, the à la Carte restaurant was located above the cafeteria with a series of smaller sitting lounges on either beam and a second cafeteria aft.

For the Danish Government, an exploitable side effect of the order for DSB was that it might be possible to stave off the closure of the struggling shipyards at Helsingør and Nakskov. For this reason, the order, placed in 1978, was split between these two yards. The new ferries, to be named *Dronning Ingrid*, *Prins Joachim* and *Kronprins Frederik* were to be delivered in 1980-81, with the former building in Helsingør and

The forward dining saloon (with panels depicting clocks by Arne L. Hansen on the bulkhead) (Kay Kørbing collection)

The 'Red Cafeteria' on the deck below: there, the clock motif was repeated in a glass partition separating the forward and aft sections. The ship's structural framing is expressed in the external walls, effectively dividing the perimeter seating into a series of booths. (Kay Kørbing collection)

the latter two in Nakskov.

The new ships – with their straight lines, and angular forms – introduced a new look to the DSB fleet. Externally, they were unapologetically functional and even the masts, funnel and deckhouses had a solid, chunky appearance. Yet, their length and repetitive window detailing made them appear also orderly and purposeful. Just as the gently curved woodwork lining the saloons of the *England* and *Sagafjord* reflected their graceful, sweeping lines, the new DSB ships' interiors reflected their exteriors through being appropriately solid and hard wearing. To create a relaxing atmosphere, off-white laminate was used for wall coverings with dark brown brushed aluminium panelling in connective spaces. Smoke-tinted glass was used for doors and earthy hues for carpeting and upholstery. These neutral tones were accented by primary red and blue being applied to such details as door handles, staircase balustrades and artworks to colour code the ships' forward and aft areas. According to Kay Kørbing:

'One problem in designing interiors for such broad, regularly-shaped ferries with relatively low deck heights compared with their width was that there were potentially vast unadorned expanses of ceiling, which could easily feel oppressive. We got over this problem very simply partly by highlighting the ship's main structural members as space dividers and partly by breaking up the ceilings with circular cut-outs with concealed perimeter lighting to give a greater sense of height in the centres of the saloons.'[53]

The bulkheads on each ship were decorated differently by prominent Danish artists

– each of whom had worked previously with Kørbing on past projects. Thus, Ole Schwalbe produced a graphic motif on six enamel panels in the ferry square of the *Prins Joachim*. These motifs then re-appeared individually throughout the ship's passenger accommodation. The *Dronning Ingrid* had perhaps the most effective scheme of the three, displaying striking panels by Arne L. Hansen. The central ferry square featured two-storey high panels, cladding the exhaust uptakes, wittily using the signal red, black and white stripes of the DSB funnel livery. The forward and aft stair halls were entirely lined with panels depicting abstractions of the ship's lines plan. The third of the trio, *Kronprins Frederik*, introduced in April 1981, featured whimsical murals of fairy tale characters and sea creatures by Helge Refn.

Able to do the work of up to seven previous ships, the new trio was a great success on the Great Belt route and, while they only marginally reduced the crossing time, their impressive range of passenger facilities at least made a welcome and relaxing break in long-distance train journeys. The DSB Inter-city ferries were the last designed by Kørbing's office which he closed in 1982. Although officially semi-retired, he enjoyed further private consultancy work, in collaboration with other architects, on several more vessels for DSB.

The entire DSB fleet was transferred to a new company, DSB Rederi A/S, in 1995 (in 1997, this firm changed its name to Scandlines). When the Great Belt Bridge opened on 1st June 1997, the three Inter-city ferries were laid up in Nyborg and Korsør. That December, the *Kronprins Frederik* was re-activated by Scandlines to serve on the Gedser-Rostock route and was rebuilt as a car ferry. Thereafter, in June 1998 the *Dronning Ingrid* and *Prins Joachim* moved to join a number of other redundant ex-DSB ferries awaiting sale in Nakskov. The *Dronning Ingrid* was sold in 1999 to the Scottish transport tycoon Ann Gloag (of Stagecoach) for conversion to a hospital ship to be run by the American Christian charity, Mercy Ships, as the *Africa Mercy*. She is presently being radically rebuilt on a piecemeal basis at the former-Cammell Laird shipyard at Hebburn-on-Tyne, near Newcastle. A new deck containing cabins has been inserted through the upper level of the former train deck (its lower level will be made into a hospital). The aft public rooms have been gutted and these too will eventually become cabins and administration offices. Completing the project will involve significant fundraising before the *Africa Mercy* can enter service as the world's largest non-military hospital ship.

In 2001, the Prins Joachim joined her sister on Scandlines' Gedser-Rostock route. Today, she is the least-altered of the three former Nyborg-Korsør Inter-city ferries.

The *Prins Joachim* in her present Scandlines livery, in service between Gedser and Rostock. (FotoFlite)

MS *NIELS KLIM* AND MS *PEDER PAARS*

The *Niels Klim* catches morning sunlight as she motors purposefully across the Kattegat from Kalundborg to Århus. The careful application of the DSB corporate identity helps to break up her bulky profile and lends an orderly, businesslike appearance. (Klas Brogren collection)

In the early-1980s, DSB's management was anxious to replace its ageing ferries on its northerly Kalundborg-Århus crossing, the longest DSB route, taking around three-hours. It was always something of a poor relation to the shorter Great Belt run and remained virtually a passenger-only route until the introduction of the first of a pair of small drive-on ferries in 1960.[54] The same ships were still operating over 20 years later and so DSB was anxious to revive the route and attempt to win back traffic lost to the slickly run shorter operation of Mols Linien between Ebeltoft and Sjællands Odde. One idea was to build three ships, each of around 10,000 gross tonnes, to offer a more regular service to compete with Mols Linien, but with greatly enhanced freight capacity. This propsal was stillborn as politics intervened. The Danish Government, on the one hand, was concerned to save the Nakskov and Helsingør shipyards from closure and, on the other, was anxious that the state-owned shipping line would not compete unfairly with private sector firms. A compromise was to order only two much larger ferries, one from each shipyard.

DSB's management decided that if it was only offered two ferries, it could not compete so effectively with Mols Linien's regular departures for passenger and car traffic and so these would best be large and relatively fast vessels with two full-height vehicle decks capable of carrying lorries and trailers with passenger accommodation above. Designs were produced, but this ambition was also curtailed to spare public money.

The final compromise, after several design proposals had been rejected, was to build two slower passenger ferries of around 20,000 gross tonnes, each with one freight deck and with an upper deck only capable of carrying cars. Both were ordered from the Nakskov Skibsværft and design work began in 1982. Following the popular

DANISH SHIP DESIGN 1936-1991

A sitting lounge on the *Peder Paars*, using classic modern lighting by Poul Henningsen, whose designs had graced the interiors of Fisker's *Kronprins Frederik* almost fifty years previously. (Dampa A/S)

Classic modern Danish furniture designs and neutral fabric tones are set off by colourful bulkhead murals in the *Peder Paars*'s cafeteria. (Dampa A/S)

The ferry square on the *Peder Paars* was an impressive space. With daylight streaming through from above and with the complete avoidance of clutter, it was also calming and relaxing. (Dampa A/S)

acclaim of his work on the Nyborg-Korsør Inter-city train ferries, Kay Kørbing was again involved in the design process, although this was mainly in an advisory role to the designers Niels Kryger and the industrial designer Christian Bjørn (the naval architect was Dwinger Marine Consult).[55]

While exclusively a passenger and road vehicle service, this route none-the-less provided important rail connections from North Jutland to Zealand and Copenhagen. Kay Kørbing recalls that:
'The new Århus-Kalundborg ships were to be by far the largest passenger ships sailing under the Danish flag in domestic service. There was the closest co-operation between the design teams and this meant that we could be much bolder in manipulating the space onboard, rather than just putting surface finishes on an architectural *fait accompli*.
'We tried many arrangements, even looking at the possibility of making double-ended ferries. That didn't happen, but as the design process advanced, we realised that, because of its width, the big jumbo ferry type had quite remarkable design possibilities. I was very pleased to be involved in the styling of the exteriors, such as the masts and funnels, as well. I have no problem accepting that such ferries are nothing like the ships of years past – except that all are built with the best technology to do their job as efficiently as possible, so we decided that if these ships were to be big floating boxes, then we would try to express this quality in such important details as the mast, funnel and the form of the superstructure.'[56]
The result of this work was certainly imposing, if a little ungainly. The ferries had

vast, slab-sided superstructures and funnels, which later proved to have an unfortunate tendency to catch side winds, occasionally making docking the ships something of a challenge for their Masters.

Clearly developed from the earlier Inter-city ferries on the Nyborg-Korsør route, these imposing ferries could take 2,000 passengers and 331 cars, or 30 18-metre trailers on the main deck and 152 cars on an upper deck, connected by internal ramps. Passenger accommodation was spread over two decks. The central focus was an impressive atrium of a scale commensurate with the size of the ships. It cut right through the main passenger decks and was topped by a series of rooflights to allow daylight to flood into the heart of the ships. To allow maximum lighting and the best views, the public rooms were all located around the perimeter of the superstructure with the circulation spaces along the ships' centre line, effectively reversing the layout of the Nyborg-Korsør ships. As they were sized to handle peak season traffic loads during the summer holiday, another good reason for this arrangement was that certain areas could be closed off altogether in winter to prevent the cavernous interiors from feeling empty and desolate. In anticipation of new fire safety regulations, the main staircases, fore and aft, were in completely sealed service towers, but again used colour coding in yellow and pink to improve passenger orientation (their ferry squares amidships were in blue). These staircases were of a novel design which showed what a fundamental re-think the design of the ships involved:

'When one sailed on a typical large Scandinavian car ferry in the holiday season and the announcement came for car drivers and passengers to go back to their cars, there was often pandemonium on the stairs, so we decided to double them on the new ships with a back-to-back arrangement, much like the famous staircase in the palace at Versailles. That split the passenger flows in two, which was much more easily controllable.'[57]

The restaurants and seating halls were to be compact and even the main cafeteria, with over 300 seats, was split into several sections. All of the passenger accommodation was finished in neutral shades of white, grey and blue with matt aluminium details. Such colour as there was came from vibrant abstract murals which liberally adorned the transverse bulkheads in the hallways, straircases and public rooms. The artworks on the *Niels Klim* were by Freddie A. Lerche, Knud Hvidberg, Lis Nogel and Gunnar Westman. Those on the *Peder Paars* were by Anders Tinsbo, Viera Collaro, Peter Hentze and Tom Krøjer. Susanne Ussing imaginatively designed the play equipment in the large childrens' playrooms.

A big effort to court business travellers was very evident with fully equipped writing rooms offering a range of facilities including typewriters, copiers, a computer terminal and pay telephones, capable of receiving incoming calls. For executives wanting more privacy, there were individual office cabins with typewriters and telephones. There were also 72 two-berth cabins with private facilities, oddly located below the car deck, which DSB attempted to utilise by offering round-trip conference packages. The interiors were furnished with specially designed fittings and modern Danish furniture classics, such as Arne Jacobsen's white bent plywood chairs, used in the cafeteria. Even upholstery and curtain textiles were specially designed and woven for the vessels.

This fundamental approach to shipboard design was in sharp contrast to that deployed on the many large jumbo ferries entering service throughout Scandinavia for other operators at that time. The Swedish interior designer, Robert Tillberg, had practically cornered the market to decorate the interiors of these ships and his standard approach relied on impact, rather than subtlety, and on surface decoration, rather than spatial manipulation. His work was heavily influenced by American-based cruise ships and featured eye-popping combinations of vibrantly-patterned carpets, mirrors, shiny brass railings, cut glass chandeliers, Tiffany lamps and ceilings, fake Gustavian furniture and Austrian festoon blinds. In contrast, the new DSB ferries were calmly understated and businesslike. Interestingly, commentary in shipping industry journals from the time of their introduction criticised their interiors for being rather too uncompromisingly severe in places and somewhat lacking in soft 'homely' comforts. DSB, however, sought to provide clean, spacious and efficient point-to-point transportation, rather than stuffiness and decorative frivolity, and it was these particular qualities that made its ferries especially distinctive.

To accommodate its new vessels, DSB's investment included the major shore works necessary in both Kalundborg and Århus. There, Mogens Steen of DSB Bygningstjenesten with Arkitektgruppen I Århus planned a suitably robust-looking new terminal building, designed to complement the profiles of the new ferries.

When handed over in October 1985 and May 1986 as *Peder Paars* and *Niels Klim*, taking names from the characters from one of the best known works of the Danish author Ludvig Holberg, the new ships broke a long sequence of royal or geographical names for major DSB ferries. Some voices were raised in protest over the suitability for a ship of the name *Peder Paars*, who, as a fictional merchant travelling to Århus, was beset with problems that prevented him from ever arriving. The criticism seemed prophetic as the new ferry had a number of hair-raising scrapes at Århus and sustained minor damage in a series of collisions with the berth. The *Niels Klim* also had her share of bumps in the early stages, the main problems arising from wind resistance to 3,200 square metres of side surface area.

Regrettably, the *Peder Paars* and *Niels Klim* never attracted the hoped-for custom as they were simply too big for the route and they could not offer as frequent departures as the rival Mols Linien and, notwithstanding their great size, they had too little freight capacity. Political interventions also cut short their careers as the Danish government at last decided to build a bridge over the Great Belt, which, upon completion, would render much of the DSB fleet redundant. At the same time, DSB wanted to order a new ship for its short high-capacity Helsingør-Helsingborg route. The Government thought that it could sell the *Niels Klim and Peder Paars* for a quick profit and give DSB some of the money back to pay for the necessary new ship. Consequently it ordered DSB to sell the modern sister ships in 1990.

The recent histories of the *Niels Klim* and *Peder Paars* have not been happy. Both ships were sold to Stena Line, having been replaced in May 1991 by the chartered converted freight ferries *Ask* and *Urd*. These ships proved to be in poor mechanical condition and initially suffered from frequent break-downs. The *Peder Paars* was re-named *Stena Invicta* and sent to the Schichau-Seebecksverft in Germany for a comprehensive rebuilding before entering service on Stena's Dover to Calais route. All traces of her original design were removed and Stena even had the atrium decked over to make more room for slot machines. Worse still, the roof lights were plated in as the British authorities claimed that the glass used in them was not safe enough. The vessel was completely ruined.

The *Stena Invicta* proved too slow for English Channel service and had only one freight deck. Consequently, she was withdrawn in 1998 when Stena merged with its former rival P&O to better compete with the Channel Tunnel. The *Stena Invicta* was then chartered to Silja Line for service between Umeå and Vaasa across the Gulf of Bothnia. The ship is now the *Color Viking* on Color Line's Sandefjord-Strömstad route.

The former *Niels Klim* was renamed *Stena Nautica*. She was fitted with stabilisers and chartered to B&I Line for Irish Sea service in 1992. Although a rather kitsch Irish-themed pub was installed, otherwise the interior remained largely intact at first. There followed a number of other short charters in the Mediterranean and Scandinavian waters. The ship now sails on Stena Line's Grenå-Varberg route, carrying freight all year round and passengers in the summer season only. In the winter of 2001-2002, the *Stena Nautica* was completely rebuilt as a ro-pax ferry at the Öresundsvarvet in Landskrona. Her lower saloon deck was dismantled to heighten the upper car deck and the height of the funnel was truncated to save weight. New interiors were designed by Figura of Gothenburg, which obliterated any remaining remnants of the original design. The result of this work is not an attractive sight.

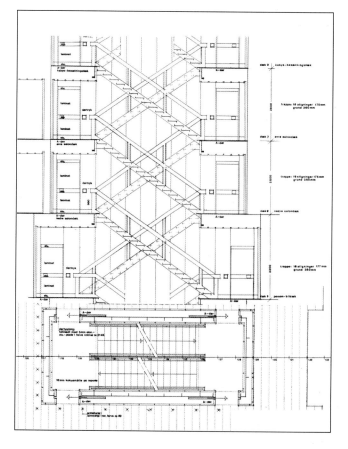

The forward yellow stair tower on the Peder Paars, showing the back-to-back stairs to double circulation and the matching wall murals. There was an identical fully-enclosed stair tower aft with a pink colour scheme. The sectional drawing of this arrangement clearly shows the orderliness of the detailing. (Dampa A/S)

OTHER PROJECTS FOR DSB

Once the *Niels Klim and Peder Paars* had entered service, DSB decided to modernise its Halskov-Knudshoved car ferry service, but by then with construction of a fixed road and rail link being planned, new ships were out of the question. Instead, it was decided to refurbish the 1973-built Halskov-Knudshoved car ferry, *Romsø* and to charter ships to replace some of the other ageing vessels on the route. Kay Kørbing designed new interiors for *Romsø* similar in style to those of his previous DSB ships, although an existing ship offered hardly any possibility of making structural changes. She was modernised over the 1986-7 winter season.

The *Romsø* was built in 1973 and its overhanging forward superstructure and angular design perhaps made it a forerunner to DSB's three Inter-city train ferries (see above). In 1986-7, the interiors were renovated by Kay Kørbing in a manner similar to those of the *Niels Klim* and *Peder Paars* (above right). Because he had to work around the existing superstructure and to a tight budget, the *Romsø* was perhaps less outstanding than other projects. (Author's collection)

The *Kraka* is seen crossing the Great Belt in 1994 – nothing, however, can disguise her origin as a perfunctory ro-ro freighter. (Author)

Next, DSB chartered three freight ferries from Per Henriksen's Mercandia Rederierne, which it intended to have rebuilt with passenger accommodation. These ships, of Mercandia's 'multiflex' design had been constructed in Frederikshavn in the early-1980s for the international charter market and were designed for long sea voyages. They had only single-screw propulsion and seemed to be patently unsuitable for the short Great Belt crossing. Interestingly, as Kørbing's firm had originally designed their crew accommodation, he was asked to prepare plans to convert them to combined passenger, car and freight carriers. It was a case of making the best of a difficult job and their new passenger saloons were attached to the rear of their existing superstructures. Little could disguise the fact that these ships were essentially cargo vessels and of rather perfunctory design. They entered service in 1988-9 as *Lodbrog, Kraka* and *Heimdal*. To improve their manoeuvrability, they were ingeniously converted to triple screw ships with their existing fixed propellers being supplemented by two azimuthing propellers, the pitch and direction of which could be altered independently.

The ships were popular with freight hauliers, however, on account of their wide and unobstructed car decks. Moreover, thanks to Kørbing's ingenious interior design work, their passenger facilities were actually remarkably bright and spacious with fresh colours and large windows throughout. They had recliner saloons, two cafeterias each and small à la carte restaurants. (In addition, the *Lodbrog* was fitted with a tax-free shop in 1989 for use on DSB's international routes).

The *Romsø* was withdrawn by Scandlines in 1998 and laid up at Nakskov. The vessel now sails in Indonesian waters as the *Agoamas*. The *Lodbrog* and *Kraka* were withdrawn at the same time. Scandlines had purchased the *Lodbrog* in 1997, but the *Kraka* was returned to Mercandia, the company from which all three had initially been chartered. The *Heimdal* was withdrawn in 1998 and she too reverted to Mercandia before being converted to a cable laying ship by Ørskov's Shipyard at Frederikshavn. The other two were converted to cable-laying ships at Cammell-Laird on the River Tyne in 1999.

MS TYCHO BRAHE

The futuristic-looking *Tycho Brahe*, seen here approaching Helsingør in 1994, brought both a higher capacity and a striking new profile to the short Helsingør-Helsingborg route. (Author)

The 20-minute long Helsingør-Helsingborg crossing was DSB's shortest international route. The fact that it was also the quickest ferry link between Sweden and Denmark made it very popular and, during the 1970s, a succession of double-ended ferries was introduced there. As the Scandinavian countries became a single trade zone as part of EFTA, more and more Swedes and Danes commuted to work using the route. DSB moved its veteran Århus-Kalundborg ships *Prinsesse Anne Marie* and *Prinsesse Elisabeth* (displaced by the jumbo ferries *Niels Klim* and *Peder Paars* described above) as a temporary measure to help cope with the seemingly insatiable demand for capacity. By the end of the 1980s, the route was served by a motley collection of Danish and Swedish ferries of many different types and sizes. Accordingly, DSB and its main competitor, Scandinavian Ferry Lines, decided to join forces and rationalise the situation by each ordering a single larger double-ended car, passenger and rail ferry, while retaining four of the older ships for rush-hour relief duties. The new joint venture between DSB and SFL was named ScandLines and it included the construction of new terminals and the introduction of a new corporate identity and ferry livery.

Notwithstanding its short length, Helsingør-Helsingborg is an attractively scenic crossing, giving fine views along the Danish and Swedish coasts up and down the Øresund and towards majestic Kronborg Castle by the entrance to Helsingør harbour. Therefore, the design of the new high-capacity ships was to feature large areas of glazing in the passenger areas. The naval architects Dwinger Marine Consult, in collaboration with DSB's technical department, worked out the technical design. Built by the Tangen Verft at Kragerø and outfitted Langsten Slip og Båtbyggeri at Tomrefjord (both in Norway), the two sisters, named *Tycho Brahe* and *Aurora af Helsingborg*, were outwardly almost identical, except that the latter had full-width enclosed navigation bridges. The new 10,845 gross tonnes near-sisters had a capacity of 1,250 passengers and 240 cars each and they appeared purposeful and decisively modern – even sculptural – in comparison with so many other new ferries of the period. As each owner had its own ideas about how the interiors should look, the Swedes entrusted this work to Robert Tillberg, while, for the last time, DSB

DANISH SHIP DESIGN 1936-1991

One of the *Tycho Brahe*'s hallways, with bright enamelled wall panels by Erik Mortensen. (Kay Kørbing collection)

Rederi employed Kay Kørbing.

Each ship had the same straightforward internal layout, featuring a large central hallway with a caferteria and a tax-free supermarket at either end on the main deck, and a restaurant, bar and fast-food kiosk with crew accommodation and a large sun deck above. On the side of the hallway facing north, there was a large double-height window to give passengers a panoramic view of Kronborg Castle and the attractive waterfront of Helsingborg. Yet, the two wildly contrasting decorative approaches were at the extremes of Scandinavian ferry interior design. The *Aurora af Helsingborg* represented Tillberg's favoured manner, with much applied ornamentation, beamed ceilings and festoon blinds at the windows – quite absurd in such a ship. In complete contrast, the Danish *Tycho Brahe* again was modern, but unpretentious and with an emphasis on light, bright, airy but comfortable spaces. According to Kay Kørbing:

'This ship really was to take a lot of punishment with potentially up to 2,500 passengers passing through every daylight hour, so it was logical to keep things simple and easily maintained. We chose high-quality tough materials, such as marble slabs for the flooring in the hallway. To make orientation easy, and to give a sense of occasion, I designed a half-spiral staircase with open treads to open up the space between the main passenger decks in the hallway. People would come onboard and, like moths, head for the daylight streaming in from the sun deck two levels above.'[58]

The artist Erik Mortensen decorated the hallways with colourful enamel panels and, once again, Kørbing designed a new tubular steel chair for use in the saloons and restaurant, while his glassfibre cafeteria chair, first used on the *Prinsesse Margrethe* of 1957, was put back into production for use in the fast food café. Having a waiter service dining saloon on such a ferry might have seemed a strange decision, but because of the fine scenery, many passengers come onboard and make several return crossings, enjoying dinner as they sail.

The *Tycho Brahe* entered service in October 1991 in the livery of ScandLines and Kay Kørbing finally retired from architecture. In 1995, with the Great Belt Bridge approaching completion, the Danish Parliament decided to split the shipping division from DSB altogether as one of the first steps in reducing DSB to be a passenger train operator only. From 1997 onwards, the majority of the former DSB ships were repainted in a revised Scandlines livery, as were the Southern Baltic ferries of the Swedish and German railways.

The *Tycho Brahe*'s cafeteria featured the same type of white glassfibre chairs as were used on his first passenger ship, the *Prinsesse Margrethe* of 1957 and they remained stylish and modern even after so many years. Each table has floor to ceiling windows, not only giving passengers a panoramic view across the sound, but also down to the ship's vehicle deck. (Kay Kørbing collection)

CONCLUSION

Arguably, Modernism in Danish architecture reached its apotheosis in the 1960s when Kay Kørbing and other architects of his generation were at the height of their powers. In the wider world, however, the modern movement was already in trouble and, indeed, appeared to be in decline. In Britain and America, its aesthetics, if not its moral and social principals, were pounced upon by property developers anxious to make a fast buck and by local authorities wishing to re-house large working class populations cheaply and quickly. Thus, all too often, modernist architecture of the post-war era became synonymous with shoddy workmanship, poor detailing and with the extensive use of unadorned concrete. Frequently, the craft ideals of Scandinavian modernism, with their emphasis on high-quality traditional materials and rational planning, were sadly forgotten and Soviet-style prefabrication and system-building were employed instead.

Simultaneously, some American architects and commentators – such as Robert Venturi – began to look to the bright lights and kitsch ornamentation of Las Vegas for new inspiration and argued for a new plurality in design, based upon signs, symbolism and decoration. Modernism was no longer the only acceptable mode of expression and it was argued that many different design tactics could be of potential merit – it was all a matter of context. Thereafter, post-modernism, as this new approach became known, dominated much architectural discourse throughout the 1970s and 80s. As it was already the language of the global leisure and hospitality industry, its impact was felt particularly strongly in cruise and ferry interior design. Thus, the emphasis changed from aspiring to produce designs capable of standing the test of time to capturing instead the spirit of the moment. Many passenger ships nowadays have interiors only intended to last for around seven to ten years. (The *Stena Danica* of Stena Line is an extreme example of this trend as she has recently been rebuilt with her fourth entirely new interior since entering service in 1983).

Although the approach favoured by Kay Fisker and Kay Kørbing has almost entirely vanished in today's highly commercial and increasingly globalised passenger shipping industry, these Danish architects have left an important legacy of innovation and, at last, the cycle of taste is coming full-circle, meaning that Danish design of the 1930s to 1970s period is once again in fashion. Moreover, subsequent ferries for the Danish part of Scandlines, such as the recent *Prinsesse Benedikte* and *Prins Richard* on the Rødby-Puttgarten route, designed by Christian Bjørn and introduced in 1997, continue 'the functional tradition' in Danish ship design.

The *Tor Britannia* is seen in her Scandinavian Seaways livery, which she carried from the mid-1980s onwards (FotoFlite)

THE WORK OF KAY FISKER AND KAY KØRBING

TECHNICAL STATISTICS

M/S HAMMERHUS
Built 1936 by Burmeister & Wain, Copenhagen, Denmark
- Yard number 622
- Dimensions .... 80,04 x 12,59 x 4,60 m
- Grt/Nrt/Dwt 1,726/ 985/ 500
- Engines .... 1st 6-cyl, Burmeister & Wain 2SA diesel
- Power output .... 2.250 hk
- Speed 15,0
- Passengers 900
- Cars 20

M/S KRONPRINS OLAV
Built 1937 av A/S Helsingørs Jernskibs og Maskinbyggeri, Helsingør, Denmark
- Yard number 246
- Dimensions .... 99,76 x 13,89 x 8,41 m
- Grt 3,038
- Engines .... Two Burmeister & Wain 750-VF-90 diesels
- Power output .... 4,800 hp
- Speed .... 18,5 knots
- Passengers 1,200

M/S C.F. TIETGEN
Built 1928 by A/S Helsingørs Jernskibs og Maskinbyggeri, Helsingør, Danmark
- Yard number .... 185.
- Dimensions .... 86,74 x 12,22 x 5,00 m
- After rebuilding .... 99,27 x 12,22 x 5,18 m
- Grt/ Nrt/ Dwt 1,850/ 1,036/ 901
- After rebuilding 1939. Grt/ Nrt/ Dwt 1,938/ 1,065/ 794
- After rebuilding 1954. Grt/ Nrt/ Dwt 2,785/ 1,652/ 935
- Engines .... One B&W 855-MTF-100 diesel
- Power output .... 1,950 hk
- Speed .... 14,7 knots
- After rebuilding 1939 .... One B&W 1050-VF-90 diesel
- Power output .... 3,600 hp
- Speed. .... 16,7 knots
- Passengers 575
- After rebuilding 1939 1105
- After rebuilding 1954 1200

M/S HANS BROGE
Built 1939 by Helsingørs Jernskibs og Maskinbyggeri, Helsingør, Denmark
- Yard number 256
- Dimensions .... 88,40 x 13,10 x 5,30 m
- After rebuilding .... 98,78 x 13,10 x 5,30 m
- Grt/ Nrt/ Dwt. 2,013/ 1,156/ 850
- After rebuilding Grt/ Nrt/ Dwt 2,927/ 1,673/ 1,050
- Engines. .... One 9-cyl, B&W 950-VF-90 diesel
- Power output .... 3,350 hp
- Speed .... 17,5 knots
- Passengers 1100
- After rebuilding 1250
- Cabin berths 239
- After rebuilding 304

M/S ROTNA
Built 1940 by Burmeister & Wain, Copenhagen, Denmark
- Yard number 655
- Dimensions .... 81,89 x 12,91 x 4,30 m
- Grt/Nrt/ Dwt 1,836/ 1,136/ 530
- Engines   One 7-cyl, Burmeister & Wain 2 SA 750-VF-90 diesel
- Power output. .... 3.000 hp
- Speed .... 15,0 knots
- Passengers 900

M/S KRONPRINS FREDERIK
Built 1941 av A/S Helsingørs Jernskibs og Maskinbyggeri Helsingør Danmark.
- Yard number .... 262.
- Dimensions .... 114,48 x 15,20 x 5,66 m
- Grt/ Dwt 3,895/1,720
- Engines .... Two 10-cyl, B&W 1050-VF-90 diesels
- Power output .... 7,100 hp
- Speed .... 20,2 knots
- Passengers 364
- Cabin berths 302

M/S KRONPRINSESSE INGRID
Built 1949 av A/S Helsingørs Jernskibs og Maskinbyggeri Helsingør Denmark
- Yard number 289
- Dimensions .... 114,48 x 15,20 x 5,66 m
- Grt 3,895
- Engines .... Two Burmeister & Wain 1050-VF-90 diesels
- Power output .... 7,100 hp
- Speed .... 20,2 knots
- Passengers 364
- Cabin berths 302

M/S KONGEDYBET
Built 1952 by Burmeister & Wain, Copenhagen, Denmark
- Yard number 685
- Dimensions .... 85,94 x 13,25 x 4,3 m
- After rebuilding .... 93,87 x 13,25 x 4,30 m
- Grt/ Nrt/ Dwt 2,314/ 386/ 565
- After rebuilding .... Grt/ Nrt/ Dwt. 2,828/ 1,653/ 640
- Engines .... One 7-cyl, Burmeister &Wain 50-VTF-110 diesel
- Power output .... 3.400 hp
- Speed .... 15,2 knots
- Passengers 1500
- Cars 24

M/S PRINSESSE MARGRETHE
Built 1957 by Helsingør Skibs & Maskinbyggeri A/S, Helsingør, Danmark
- Yard number 322
- Dimensions .... 121,03 x 16,18 x 4,87 m
- Grt/ Dwt 5,061/ 994
- Engines .... Two 8-cyl, Burmeister & Wain diesels
- Power output .... 7,300 hp
- Speed .... 20,5 knots
- Passengers 1200
- Cabin berths 395
- Cars 35

M/S KONG OLAV V
Built 1961 by Aalborg Værft A/S, Aalborg, Denmark
- Yard number 135
- Dimensions .... 121,01 x 16,18 x 5,10 m
- Grt/ Nrt/ Dwt 4,555/ 2,283/ 1,135
- Engines .... Two 8-cyl, Burmeister & Wain diesels
- Power output .... 7,500 hp
- Speed .... 20,5 knots
- Passengers 977
- Cabin berths 220
- Cars 18

DANISH SHIP DESIGN 1936-1991

M/S ENGLAND
Built 1964 by Helsingør Skibs & Maskinbyggeri A/S, Helsingør, Denmark
- Yard number 369
- Dimensions .... 140,00 x 19,33 x 5,54 m
- Grt/ Dwt 8,221/ 1,451
- Engines .... Two 10-cyl, B&W 1050-VT2BF-110 diesels
- Power output .... 14,000 hp
- Speed .... 21,0 knots
- Passengers 467
- Cabin berths 467
- Cars 100
- After rebuilding 120

M/S SAGAFJORD
Built 1965 by Forges et Chantiers de la Méditerranée, La Seyne, France
- Yard number 1366
- Dimensions .... 188,88 x 24,46 x 8,30 m
- Grt/ Dwt 24002/ 6353
- Engines .... Two 9-cyl, Sulzer diesels
- Power output .... 17650 kW
- Speed .... 20,0 knots
- Passengers 789
- Cabin berths 789

M/S WINSTON CHURCHILL
Built 1967 by Cantieri Navali del Tirreno e Riuniti S.P.A. Riva Trigoso, Genoa, Italy
- Yard number 277
- Dimensions .... 140,65 x 20,53 x 5,59 m
- Grt/ Nrt/ Dwt 8,657/ 4,778/ 2,235
- Engines ... Two 2D 2SA 10-cyl, B&W 1050-VT2BF-110 diesels
- Power output .... 14,000 hp
- Speed .... 21,0 knots
- Passengers 450
- After rebuilding 590
- Cabin berths 390
- Cars 180

M/S KONG OLAV V
Built 1968 by Cantieri Navali del Tirreno e Riuniti S.P.A. Riva Trigoso, Genoa, Italy
- Yard number 278
- Dimensions .... 124,95 x 19,28 x 5,22 m
- Grt/ Nrt/ Dwt 7,965/ 3,912/ 1,108
- After rebuilding .... Grt/ Nrt/ Dwt. Brt. 8,669/ 4,441/ 1,084
- Engines .... Two 12-cyl, B&W 1242-VT2BF-90 diesels
- Power output .... 12,000 hp
- Speed .... 21,0 knots
- Passengers 952
- After rebuilding 1100
- Cabin berths 506
- After rebuilding. 698
- Cars 100
- After rebuilding 121

M/S PRINSESSE MARGRETHE
Built 1968 by Cantieri Navali del Tirreno e Riuniti S.P.A. Riva Trigoso, Genoa, Italy
- Yard number 279
- Dimensions .... 124,95 x 19,28 x 5,22 m
- Grt/ Nrt/ Dwt 7,965/3,912/ 1,100
- After rebuilding .... Grt/ Nrt/ Dwt. 8,669/ 4,441/1,084
- Engines .... Two 12-cyl, B&W 1242-VT2BF-90 diesels
- Power output .... 12,000 hp
- Speed .... 21,0 knots
- Passengers 952
- After rebuilding 1100
- Cabin berths 506
- After rebuilding 698
- Cars 100
- After rebuilding 121

M/S AALBORGHUS
Built 1969 by Cantieri Navali del Tirreno e Riuniti S.P.A. Riva Trigoso, Genoa, Italy
- Yard number 280
- Dimensions .... 124,85 x 19,31 x 5,21 m
- Grt/ Nrt/ Dwt 7,697/ 3,890/ 1,021
- After rebuilding .... Grt/ Nrt/ Dwt. 7,988/ 3,669/ 975
- Engines .... Two 12-cyl, B&W 1242-VT2BF-90 diesels
- Power output .... 12,000 hp
- Speed .... 21,0 knots
- Passengers 718
- After rebuilding 691
- Cabin berths 506
- After rebuilding 691
- Cars 120

M/S TREKRONER
Built 1970 by Cantieri Navali del Tirreno e Riuniti S.P.A. Riva Trigoso, Genoa, Italy
- Yard number 281
- Dimensions .... 124,85 x 19,31 x 5,21 m
- Grt/ Nrt/ Dwt 7,692/ 3,669/ 1,125
- Engines .... Two 12-cyl, B&W 1242-VT2BF-90 diesels
- Power output .... 12,000 hp
- Speed .... 21,0 knots
- Passengers 718
- Cabin berths 506
- Cars 120

M/S VISTAFJORD
Built 1973 by Swan Hunter Shipbuilders, Wallsend, Great Britain
- Yard number 39
- Dimensions .... 191,09 x 25,00 x 8,20 m
- Grt/ Dwt 24292/ 5954
- Engines. .... Two 9-cyl, Sulzer-Clark diesels
- Power output .... 17650 kW
- Speed. .... 20,0 knots
- Passengers. .... 500. (max 670)
- Cabin berths .... 500 lower berths (670 including upper berths)

M/S DANA REGINA
Built 1974 by Aalborg Værft A/S, Aalborg, Denmark
Yard number 200
Dimensions .... 153,7 x 22,7 x 6,00 m
Grt/ Nrt/ Dwt 10,002/ 6,158/ 2,850
Engines .... Four 8-cyl, Burmeister & Wain diesels
Power output .... 12,945 kW
Speed .... 18,0 knots
Passengers 1065
Later increased to 1500
Cabin berths 861
Cars 300
After rebuilding 370
Freight metres 600

M/S TOR BRITANNIA
Built 1975 by Lübecker Flender-Werke, Lübeck, Germany
Yard number 607
Dimensions .... 182,26 x 23,62 x 6,20 m
After rebuilding .... 184,55 x 26,4 x 6,25 m
Grt/ Nrt/ Dwt 15,657/ 7,729/ 3,200
After rebuilding. .... Grt/ Nrt/ Dwt. 21,545/ 12,052/ 3,335
After rebuilding 2000 .... Grt, 22 528
Engines .... Four 12-cyl, Pielstick diesels
Power output .... 45,600 hp
Speed .... 27,2 knots
Passengers 1507
Cabin berths 1,234
After rebuilding .... Cabin berths. 1,416
Present capacity 1519
Cars 420
Freight metres 910

M/S TOR SCANDINAVIA
Built 1976 by Lübecker Flender-Werke, Lübeck, Germany
Yard number 608
Dimensions .... 182,26 x 23,62 x 6,20 m
After rebuilding .... 184,55 x 26,4 x 6,20 m
Grt/ Nrt/ Dwt. 15,673/ 7,756/ 3,290
After rebuilding .... Grt/ Nrt/ Dwt. 21,545/ 12,099/ 3,335
Engines .... Four 12-cyl, Pielstick PC3 12 V480 diesels
Power output .... 45,600 hp
Speed .... 27,2 knots
Passengers 1,507
Cabin berths 1,416
After rebuilding 1,617
Cars 420
Freight metres. 910

M/S DRONNING INGRID
Built 1980 by Helsingør Skibs & Maskinbyggen A/S Helsingør Denmark
Yard number 418
Dimensions .... 152,00 x 23,70 x 5,64 m
Grt/ Nrt/ Dwt 10,606/ 5,088/ 5,199
Engines .... Six 16-cyl, B&W Alpha 16U28LU diesels
Power output .... 25,440 hp
Speed .... 18,5 knots
Passengers 2280
Total railway tracks 4
Freight wagons 60
Total usable track length .... 494,0 metres
Cars. 200
Freight metres 625

M/S PRINS JOACHIM
Built1980 by Nakskov Skibsværft A/S Nakskov Denmark
Yard number 223
Dimensions .... 152,00 x 23,12 x 5,60 m
Grt/ Nrt/ Dwt 10,616/ 5,088/ 2,300
After rebuilding .... Grt/ Nrt/ Dwt. 16,071/ 4,821/ 4,500
Engines .... Six 16-cyl, B&W 16U28LU diesels
Power output .... 17,200 kW
Speed .... 18,5 knots
Passengers 2,300
After rebuilding 1400
Cars. 200
After rebuilding 210
Freight metres. 625
Total railway tracks 4
Freight wagons 60
Total usable track length .... 494,0 metres
Cars 200
Freight metres 625

M/S KRONPRINS FREDERIK
Built 1981 by Nakskov Skibsværft A/S, Nakskov, Denmark
Yard number 224
Dimensions .... 152,00 x 23,70 x 6,00 m
After rebuilding .... 152,00 x 23,70 x 5,10 m
Grt/ Nrt/ Dwt 10,616/ 5,088/ 5,139
After rebuilding .... Grt/ Nrt/ Dwt. 16.071/ 4.821/ 4.500
Engines .... Six 16-cyl, B&W 16U28LU diesels
Power output .... 25,400 hp
Speed .... 18,5 knots
Passengers 2280
After rebuilding 1400
Cars 180
After rebuilding 225
Freight metres 494
After rebuilding 724
Total railway tracks 4
Freight wagons 60
Total usable track length .... 494,0 metres
Cars 200
Freight metres 625

M/S PEDER PAARS
Built 1985 by Nakskov Skibsværft Denmark
Yard number 233
Dimensions .... 134,02 x 24,61 x 5,65 m
After rebuilding .... 137,00 x 24,61 x 5,65 m
Grt/ Nrt/ Dwt 11,602/ 6,180/ 2,813
After rebuilding .... Grt/ Nrt/ Dwt. 19,763/ 9,130/ 2,238
Engines. .... Two B&W 8L45GB diesels
Power output .... 16,960 hp
Speed .... 17,5 knots
Passengers 2000
After rebuilding 1720
Cabin berths 148
Cars 350

M/S NIELS KLIM
Built 1986 by Nakskov Skibsværft Denmark
Yard number 234
Dimensions .... 134,0 x 24,6 x 5,7 m
After rebuilding .... 135,47 x 24,6 x 5,7 m
Grt/ Nrt/ Dwt 11,763/ 6,154/ 2,813
After rebuilding .... Grt/ Nrt. 19,504/ 6,180
Engines .... Two B&W 8L45GB diesels
Power output .... 12,470 kW
Speed .... 17,5 knots
Passengers 2000
After rebuilding 663
Cabin berths 148
Cars 411

M/S ROMSØ
Built 1973 by Helsingør Skips & Maskinbyggeri A/S Helsingør Denmark
Yard number 402
Dimensions .... 130,00 x 17,70 x 5,00 m
Grt/ Nrt/ Dwt 5,603/ 2,436/ 1,604
Engines .... Two B&W DM 10U45 HU diesels
Power output .... 12,000 hp
Speed .... 18,0 knots
Passengers .. 1200 in winter service. 1500 in summer service
Cars 338
Freight metres 365

M/S KRAKA
Built 1982 by Frederikshavn Værft A/S, Frederikshavn, Denmark
Yard number 402
Dimensions .... 131,70 x 20,00 x 4,66 m
Grt/ Nrt/ Dwt 3,041/ 1,713/ 7,200
After rebuilding .... Grt/ Nrt/ Dwt. 4,290/ 2,444/ 2,250
Engines .... One MaK 12M453AK diesel
Power output .... 5,000 hp
Speed .... 15,0 knots
Passengers 12
After rebuilding 540
After rebuilding .... Cars. 275

M/S LODBROG
Built 1982 by Frederikshavn, Værft A/S, Frederikshavn, Denmark
Yard number 401
Dimensions .... 131,70 x 20,11 x 6,16 m
After rebuilding .... 141,50 x 20,11 x 6,16
Grt/ Nrt/ Dwt 3,041/ 1,713/ 7,200
After rebuilding. Grt/ Nrt/ Dwt 10,404/ 3,122/ 2,400
After rebuilding. Grt 10,068
Engines .... One MaK 12M453AK diesel
Power output .... 5,000 hp
Speed .... 15,0 knots
After rebuilding. Passengers 500
After rebuilding. Cars 290

M/S HEIMDAL
Built 1982 by Frederikshavn Værft A/S, Frederikshavn, Denmark
Yard number 409
Length .... 131,70 x 20,00 x 4,75 m
Grt/ Nrt/ Tdw 3,041/ 1,713/ 7,200
After rebuilding. Grt/ Nrt/ Dwt 9,975/ 2,992/ 2,256
Engines .... One MaK 12M453AK diesel
Power output .... 5,000 hp
Rebuilt with three Caterpillar-generators of 2,000 hp coupled to two Azimuthing propellers for better manoeuvring in harbours
Speed .... 15,0 knots
Passengers after rebuilding 540
Cars after rebuilding 275
Freight metres 891

M/S TYCHO BRAHE
Built 1991 by Tangen Verft A/S, Kragerø, Norway. Outfitted by Langsten Slip & Båtbyggeri, Tomrefjorden, Norway.
Yard number 156
Dimensions .... 110,20 x 28,20 x 5,50 m
Grt/ Nrt/ Dwt 10,845/ 3,253/ 3,060
Engines .... Four Wärtsilä-Vasa 6R32E diesels
Power output .... 9,840 kW
Speed .... 13,5 knots
Ice class 1C
Passengers 1250
Cars 240
Track Length .... 260 metres

Notes

1 The Royal Academy of Fine Arts (Det Kongelige Akademie for de Skønne Kunster, colloquially Kunstakademiet) was founded by King Frederik V in 1754 and has throughout its history occupied Charlottenborg Palace in Copenhagen. Its immediate object as established was to train the artists so greatly needed by the absolute monarchy, but in course of time, the Academy has broadened its policy and is now perhaps the most respected college of architecture, painting, and sculpture in Denmark. From its foundations, the Academy has had two main functions. Through the Council of Fellows, it has been the official adviser to the Government on matters of art; and through its schools it has controlled the teaching of the three arts.

2 One of the most influential architectural theorists of the 19th century, Eugene Emmanuel Viollet-le-Duc impacted not only French architecture, but also work in England and especially in America. His writing encouraged a great deal of creative thought and debate regarding honest structural expression and the embracing of modern technology. Henry Van Brunt's translation of Viollet-le-Duc's "Discourses on Architecture" was published in 1875, making it available to an American audience little more than a decade after its initial publication. In addition to his written work, Viollet-le-Duc is remembered for his restoration of numerous French cathedrals, including Notre Dame.

3 Le Corbusier Vers Une Architecture

5 Interview with the author 14.7.02

6 Notes for an article by Poul Kjærgaard in Arkitektura 15 (1993) kindly supplied by Pauli Wulff of Poul Kjaergaard Architects, Copenhagen

7 Ibid

8 Ibid

9 BornholmsTrafikken 1866-1991 p18.

10 Op cit Kjaergaard

11 Ibid

12 Ibid

13 Ibid

14 Ibid

15 Ibid.

16 Ibid

17 Ibid

18 So far as can be ascertained, Gerasimos Ventouris was not directly related to the Ventouris family of the Ventouris Ferries, Ventouris Sea Lines and A.K. Ventouris companies. All, however, came from the island of Kimolos.

19 Because of the decline in tourism to Bornholm, the 66 Company sold its motorship *Bornholm* to Det Nordenfjeldske Dampskibselskab in 1940.Later, the steamer *Vesterhavet* struck a mine while carrying German prisoners of war in June 1945. See Bornholmstrafikken 1866-1991

20 Op cit Kjaergaard

21 Comment by Keld Helmer Petersen to the author, 21.7.01

22 This Asplund was not related to Eric Gunnar Asplund.

23 Interview with Kay Kørbing by the author 6th January 2000

24 Ibid

25 Ibid

26 Ibid

27 *MS Prinsesse Margrethe* (DFDS, Copenhagen, May 1957)

28 His assistants included Knud Rasmussen, Bøge Jensen, Paul Meriluoto and Helmer Hansen.

29 Interview with Bryan Corner-Walker

30 Interview with Kay Kørbing by the author 6th January 2000

31 All of Kay Kørbing's furniture designs were code-named with a 'KK' prefix, followed by a number.

32 Ibid

33 Letter from Dr Andrea Ginnante to the author, dated 4th May 2000.

34 Interview with Kay Kørbing by the author 6th January 2000.

35 The Aalborghus was named after Aalborg Castle, but the Trekroner took its name from the 'three crowns' symbol under which the three kingdoms of Scandinavia were amalgamated in the Union of Kalmar (1397-1523) during the reign of Queen Margrethe I.

36 Interview with Kay Kørbing by the author 6th January 2000

37 SOLAS – the international Safety Of Life At Sea convention. These statutory rules demanded a more fundamental subdivision of ships into vertical and horizontal fire zones, separation of passenger spaces from areas of high fire risk and the provision of adequate and protected fire escape routes. The new standards also required greater use of incombustible and low flame spread materials, also higher standards for insulation, fire doors and other details of construction. The ingenuity of the designer in

producing a pleasant environment whilst working with a severely limited choice of materials and fittings was thus stretched further.
38 From the *Vistafjord*'s inaugural brochure.
39 These were the *Freeport*, *Golden Odyssey*, *Gotland* and *Visby*.
40 Interview with Kay Kørbing by the author 6th January 2000
41 Interview with Bryan Corner-Walker, former DFDS Technical Director.
42 Ibid.
43 It has been suggested that the order nearly went instead to the Dubegion Normandie Shipyard at Nantes in France, which had recently completed three modern ferries for the constituent companies of Silja Line – the *Bore Star*, *Wellamo* and *Svea Corona*.
44 Interview with Thomas Wigforss, presently of Cenargo Shipping.
45 Ibid.
46 Information from Klas Brogren
47 Interview with Kay Kørbing..
48 Interview with Thomas Wigforss
49 These were the *Stena Atlantica*, *Stena Nautica*, *Stena Nordica* and *Stena Normandica*, built by Rickmers Werft in Bremerhaven, West Germany in 1974-5.
50 Assistants working with Kørbing and Kryger were Kurt Eriksen, Carsten Dall, Ida Madsen-Mygdal, and Kimmi Dejgaard. The internal signage was designed by Sten Lange and Jørn Damgaard.
51 Interview with Kay Kørbing by the author 19th June 2001
52 Ibid
53 Ibid
54 This was the *Prinsesse Anne Marie*. She was followed by a similar sistership, the *Prinsesse Elizabeth*, in 1964.
55 Assistants working with Kørbing, Kryger and Bjørn were Hanne Flarup, Helle Borten, Kimmi Dejgaard and Jesper Greiffenberg. Signage was designed by Steen Lange and Jørn Damsgaard.
56 Interview with Kay Kørbing by the author 19th June 2001
57 Ibid
58 Interview with Kay Kørbing by the author 19th June 2001

BIBLIOGRAPHY
BOOKS
Brummer, Carl Mennesker, Huse- og Hunde (Copenhagen, 1949) pp 80-82 and 95-100
Det Forenede Dampskibs Selskab Fem Aars Genopbygning (DFDS Copenhagen, December 1950)
Frigaard, Anne Marie and Jacobsen, Gert Bornholmstrafikken 1866-1991 (Rønne, 1991)
Kraks' Blå Bog 1999
Thorsøe, Søren, Simonsen, Peter, Krogh-Andersen, Søren, Fredrichsen, Frederik and Vaupel, Henrik DFDS 1866-1991: Skibsudvikling Gennem 125 År (Copenhagen, 1991)

PERIODICALS
The Architectural Review
602 February 1947 pp51-56 *Kay Fisker: MS Kronprins* Frederik
Arkitekten
1914 pp49-54 Carl Brummer: *Skibsinteriører*
1936 pp158-60 Ole Wanscher: *Indretningen af »Hammershus«*
1949 pp106-111 Kay Fisker: *Skibsaptering*
1960 pp 59-62 Hans Erling Langkilde: *Arkitekten Kay Fisker*
Arkitektura
No 15 1993 Poul Kjærgaard: *Kay Fisker Til Søs*
Arkitektur DK

1958 No 1 pages 12-17 Poul Erik Skriver: *DFDS Offices, Århus*
1962 No 6 pages 78-84 Poul Erik Skriver: *Illums Bolighus Store, Copenhagen*
1957 No 5 pages 137-147 Bent Moudt: *DFDS' Ruteskib MS »Prinsesse Margrethe«*
1981 No 2 pp45-51 Jørgen Sestoft: *Form på Rejsen*
1981 No 2 pp52-61 *Nye DSB-færger på Storebælt*
1986 No 3 pp100-109 Poul Erik Skriver: *Kalundborg-Århus færgerne*
DSB Bladet
July 1980 pp12-14 Kay Kørbing: *En Markant Færgeprofil*
Mobilia
No 77 December 1961 *Illums Bolighus ombygget af arkitekt Kay Kørbing*
Nos 110-111 September-October 1964 *MS England*
The Motor Ship
November 1974 pp1225-1234 *The Luxury Passenger/Car Ferry 'Dana Regina'*
Nyt Tidskrift for Kunstindustri
1936 pp105-109 *Motorskibet »Hammershus«*
Shipbuilding and Shipping Record
June 11 1964 p778 *The Danish-built and –owned "England"*
June 29 1967 p901 *'Winston Churchill'*
August 2 1968 p159 *'Kong Olav V'*

BROCHURES
DFDS Præsenterer MS Trekroner (Bendt Wikke, Marketing)
MS Prinsesse Margrethe (DFDS, Copenhagen, May 1957)

ACKNOWLEDGEMENTS

Aalborg University Library, Micke Asklander, Anders Bergenek, Jens Bertelsen, Mette Bliggaard at the Danish Cultural Institute, Edinburgh, Mr & Mrs Ernst Botfeldt, Klas Brogren of Shippax, David Buri at Glasgow School of Art Library, Anthony Cooke, Bryan Corner-Walker, Miles Cowsill, John F. Hendy, Dampa A/S, Esbjerg Municipal Archive and Museum, Dr Andrea Ginnante, Finncantieri, Riva Trigoso, Italy, Stephen Gooch, Ambrose Greenway, Knud E. Hansen A/S, Copenhagen, Denmark, Ole Hansen, DSB Communication Department, Søren Lund Hviid, Per Jensen, Poul Kjærgaard, Peter Knego, Bård Kolltveit, Niels Kryger, Kay Kørbing, Mick Lindsay, Bo Løser of Dampa DK, Georg and Øystein Matre at Westcon Shipyard, Ølen, Norway, Inge Mitchell, R., The Mitchell Library, Glasgow, Flemming Nielsen at Aalborg Stadsarkiv, Aalborg, Denmark, Thomas Nørregaard Olesen, John Peter, Keld Helmer Petersen, René Taudal Poulsen, Scandlines AS, Louis Schnabenborg, Ted Scull, Richard Seville, Tony Smith of the World Ship Society, Selskabet for Arkitekturhistorie, Copenhagen, Denmark, Erik Christian Sørensen, Søren Thorsøe, Henrik Vaupel, DFDS, Copenhagen, Thomas Wigforss, Pauli Wulff of Paul Kjærgaard Architects.

INDEX

Aagaard-Andersen, Gunnar, 47
Aalborg Værft, 49, 72, 80, 84
Aalborghus, 75, 76, 77, 78, 79, 114
Africa Mercy, 100
Agger, Margrethe, 86
Århus University, 5, 16
Arkitektgruppen I Århus, 105
Arneberg, Arnstein, 60
Arnoldi, Per, 72, 86, 91
Art Deco, 11
Aurora af Helsingborg, 109, 110
Bergensfjord, 59
Bjørn, Christian, 103, 112
Black Watch, 31
Bornholmstrafikken, 95
Brogren, Klas, 89, 90
Brummer, Carl, 9, 17, 25
Burmeister and Wain (B&W), 9, 15, 30, 38
C.F. Tietgen, 25, 26, 27, 113
Cantieri Navali del Tirreno e Riuniti, (Shipyard) Riva Trigoso, 66
Carnival Corporation, 82
Caronia, 83
Christensen, Markan, 86
Clausen, Erik, 86
Collaro, Viera, 104
Color Viking, 106
Corbusier, Le, 10, 11
Corner-Walker, Bryan, 51, 52, 53, 66, 67, 84, 85, 87
Costanzi, Nicolò, 66
Cunard, 10, 18, 56, 65, 82, 83, 91
Dalsgaard, Sven, 47
Dampskibs-Selskabet af 1866 paa Bornholm, 15, 28, 38
Dana Corona, 76, 78, 79, 80
Dana Regina, 3, 56, 84, 86, 87, 88, 90, 115
Dana Sirena, 76, 80
Det Forenede Dampskibs Selskab (DFDS), 3, 9, 20, 21, 22, 23, 24, 25, 26, 27, 31, 33, 35, 37, 41, 42, 43, 44, 48, 49, 50, 51, 53, 55, 57, 66, 67, 68, 70, 71, 72, 75, 76, 80, 84, 85, 87, 90, 93
Ditzel, Nanna, 58
Dronning Ingrid, 95, 96, 97, 99, 100, 115
DSB (Danske Statsbaner/Danish State Railways), 75, 94, 95, 97, 98, 99, 100, 101, 104, 105, 107, 108, 109, 111
Dwinger Marine Consult, 103, 109
Eide, Njål, 62
Ekselius, Jan, 86
England, 2, 3, 4, 13, 50, 51, 52, 53, 54, 55, 57, 59, 62, 63, 66, 67, 68, 70, 71, 84, 87, 88, 99, 114
Evensen, Phyllis, 64
Exposition Internationale des Arts Décoratifs, 11
Fisker, Kay, 1, 3, 4, 5, 8, 11, 12, 13, 14, 15, 20, 35, 38, 41, 42, 43, 44, 112
Flender Werke (Shipyard), 88
France, 64
Furniture, 57
Goldman, Morris, 57
Grönvall, Lisa, 91
Grundtvig, Ernst, 91
Gunnarson, Carl B., 63, 82
H.P. Prior, 41
Hammershus, 15, 17, 18, 19, 20, 22, 28, 30, 38, 113
Hans Broge, 25, 26, 27, 113
Hansen, Arne L., 55, 97, 98, 100
Hansen, Knud E., 20, 21, 41, 84, 89
Haug, Kaare, 60, 65
Heimdal, 108
Helmer Petersen, Keld, 40, 47, 70
Helsingør Skibsværft, 20, 25, 28, 32, 99, 101
Henningsen, Poul, 33
Henrik Gerner, 19
Hentze, Peter, 104
Hoegh, Leif, 65
Holden, Cliff, 91
Holmshov, Helge, 47
Honore, Lise, 81
Hvidberg, Knud, 104
Ile De France, 11
Jens Bang, 41
Jens Kofoed, 95
Jensen, Jens Urup, 70
JG Furniture, 57, 58
Johanson, Cyrillus, 43
Kjærgaard, Poul, 15, 16, 31, 32, 38
Knudtzon, Anne Lise, 81
Kong Olav V (I), 49, 114
Kong Olav V (II), 14, 72, 73, 74, 76, 114
Kongedybet, 38, 39, 40, 113
Kørbing, J.A., 20, 43, 51, 55
Kørbing, Kay, 1, 3, 4, 5, 8, 13, 14, 40, 41, 42, 43, 45, 53, 54, 55, 57, 62, 66, 72, 76, 81, 84, 86, 87, 88, 90, 91, 92, 94, 95, 99, 103, 107, 110, 111, 112
Krafft, Tove, 82
Kraka, 108
Krøjer, Tom, 104
Kronprins Frederik (I), 31, 32, 33, 34, 35, 36, 37, 44, 50, 53, 113
Kronprins Frederik (II), 95, 99, 100, 115
Kronprins Olav, 1, 20, 21, 22, 23, 24, 25, 26, 28, 31, 33, 35, 113
Kronprinsesse Ingrid, 31, 35, 44, 50, 113
Kryger, Niels, 95, 97, 103
Kungsholm (1928), 11
Laakso, Vuokko, 90, 91

Langsten Slip og Båtbyggeri (Shipyard), Tomrefjord, 109
Latsis, John S., 37, 56
Lauritzen, J., 4, 55
Leonardo Da Vinci, 64
Lerche, Freddie A., 104
Lodbrog, 108
Lund, Thorkil, 15, 16, 17, 38
Lyfa (lighting), 55
Melodia, Giò, 66
Mercandia Rederierne, 108
Middelboe, Rolf, 55, 86
Mies van der Rohe, Ludwig, 40
Modernism/modern movement, 4, 10, 12, 112
Møller, C.F., 5, 15
Mols Linien, 101, 105
Moltke, Ivar, 90
Mortensen, Erik, 110
Munster, 31
Nakskov Skibsværft, 99, 101
Naxos, 43
Nemon, Oscar, 70
Niels Klim, 101, 102, 104, 105, 106, 107, 109, 116
Nilsson, Finn, 62
Nilsson, Hans, 92
Nogel, Lis, 104
Norwegian America Line, 59, 60, 65, 81, 84
Ogaard, Suzanne, 82
Ohio, 43
Oklahoma, 43
Örrefors, 55, 91
Oslofjord, 59
Pearl of Scandinavia, 3
Peder Paars, 101, 102, 104, 105, 107, 109, 116
Petersen, Godtfred H., 57, 58
Peynet, Georges, 62
Platou, Fritjof S., 62
Post-modernism, 112
Povl Anker, 95
Prince of Scandinavia, 93
Princess of Scandinavia, 93
Prins Joachim, 95, 99, 100, 115
Prins Richard, 112
Prinsesse Benedikte, 112
Prinsesse Margrethe (I), 44, 45, 46, 47, 48, 49, 111, 113
Prinsesse Margrethe (II), 14, 72, 73, 74, 114
Prinsessen, 49
Queen Elizabeth 2, 86, 91
Refn, Helge, 100
Romsø, 107, 108
Rotna, 28, 29, 30, 38, 113
Royal Academy of the Fine Arts, The, 4, 43
Rumohr, Knut, 81
Saga Rose, 65, 83
Sagafjord, 59, 60, 61, 62, 64, 65, 81, 82, 91, 99, 114
Salén, Christer, 88, 89, 91, 92, 93
Scandinavian Ferry Lines (SFL), 109
Scandlines, 100, 108, 109, 111, 112
Schwalbe, Ole, 47, 100
Selandia, 9
Société des Forges et Chantiers de la Méditerranée, La Seyne, 60
SOLAS (International Safety of Life At Sea) Convention 1974, 81, 84, 93
Stena Danica, 112
Stena Invicta, 105
Stena Line, 94, 105, 106, 112
Stena Nautica, 106
Stockholm Exhibition, 12
Storä Essingen Kyrka, 43
Suecia, 31
Suenson, Palle, 41, 42
Swan Hunter (Shipbuilders) Ltd, Newcastle, 81
Terminal Buildings, 57
Thorballs Eftf, 55
Tillberg, Robert, 93, 104, 109, 110
Tinsbo, Anders, 104
Tor Britannia, 3, 88, 90, 91, 92, 93, 115
Tor Scandinavia, 88, 90, 91, 92, 93, 115
Törneman, Ulf, 91
Trafalgar House, 65
Trekroner, 75, 76, 77, 78, 79, 114
Tycho Brahe, 109, 110, 111, 116
Ussing, Susanne, 104
Uthaug, Jorleif, 82
Vana Tallinn, 87
VanTienhoven, Han, 62
Vega, 31
Venus, 21, 41
Vistafjord, 6, 7, 81, 82, 83, 84, 86, 91, 115
Wärtsilä, 85
Werenskiold, Dagfin, 63
Werner, Gösta, 91
Westman, Gunnar, 104
Wigforss, Thomas, 88, 91, 92
Wikander, Lars, 90
Winston Churchill, 3, 13, 67, 68, 69, 70, 71, 72, 84, 85, 87, 114

DANISH SHIP DESIGN 1936-1991

The *Sagafjord* shows off her sleek lines as she glides through the English Channel – a classic image of one of the most graceful Scandinavian liners and a ship renowned for her perfectionist onboard service and elegant interiors. (FotoFlite)

The semi-circular first class Garden Lounge, located forward on the saloon deck, was one of Kay Kørbing's finest ship interiors. Rich in colour, it featured a circular white marble dance floor as its centrepiece with a concealed colour-change lighting system. Around this, there were vertical slatted screens with wicker furniture and plants in troughs. (Keld Helmer Petersen)

THE WORK OF KAY FISKER AND KAY KØRBING

The *Winston Churchill*'s colourful nightclub had a metal dance floor and a cocktail bar in front of a glazed aft bulkhead. The fashions worn by the models posed for this DFDS publicity photograph are also noteworthy – especially the gentleman seated to the right, wearing sunshades. (Author's collection)

Immaculate white linen table settings contrast with rich wood veneers and Urup Jensen's abstract tapestry panel in the *Winston Churchill*'s first class dining saloon. (Author's collection)

DANISH SHIP DESIGN 1936-1991

The *Tor Britannia* is seen off Harwich in the early 1980s, painted in the original Tor Line livery. Kay Kørbing's demarcation of the white superstructure and blue hull emphasised the length of the ship and gave her a powerful forward-thrusting appearance. (Ambrose Greenway)

Moving into the 1980s, Kay Kørbing's work for DSB was influenced by Jens Nielsen and Niels Kryger. The ferry square on the *Dronning Ingrid* shows the crisp, colourful and spacious result of this collaboration. (Kay Kørbing collection)

The *Prince of Scandinavia* (ex *Tor Britannia*) shows her tiers of sun decks to the rear of the superstructure. Designed from sketches by Kay Kørbing, this layout proved to be highly popular with passengers. Forward of the funnel, there is a sheltered lido area with a swimming pool, which could only be used in calm Summer weather. Even so, this feature helped to emphasise the ship's cruise-style onboard atmosphere in Tor Line and DFDS publicity material. (FotoFlite)

DANISH SHIP DESIGN 1936-1991

The *Peder Paars'* ferry square contained her blue amidships stairwell (there were red and yellow staircases forward and aft). Clean lines, hard wearing finishes and simple, yet comfortable, furnishings made this an attractive shipboard space. (Dampa A/S)

Bulky but businesslike, the *Peder Paars* ploughs her way between Århus and Kalundborg through a choppy Kattegat. In such conditions, her large funnel and mast made her rather difficult to handle when within the confines of harbours. (Klas Brogren collection)

DANISH SHIP DESIGN 1936-1991

The *Tycho Brahe*'s spacious marble-floored hallway features an impressive spiral staircase. Light and shadow strikes across the floor from the large window on the ship's north side, facing Kronborg Castle. (Kay Kørbing collection)